# Stochastic Analysis and Applications

*Proceedings of the 1989 Lisbon Conference*

A.B. Cruzeiro
J.C. Zambrini
Editors

1991

Birkhäuser
Boston · Basel · Berlin

Ana Bela Cruzeiro
Instituto de Fisica e Matematica
  (INIC)
1699 Lisboa Codex, Portugal

Jean Claude Zambrini
Department of Mathematics
Royal Institute of Technology
S-10044 Stockholm, Sweden

**Library of Congress Cataloging-in-Publication:Data**

Stochastic analysis and applications : proceedings of the 1989 Lisbon
  conference / A.B. Cruzeiro, J.C. Zambrini, editors.
        p.    cm.  --  (Progress in probability : v. 26)
   Papers from the First Lisbon Conference on Stochastic Analysis and
   Applications, held Sept. 1989.
       Includes bibliographical references.
       ISBN 0-8176-3567-X (hard : acid-free) -- ISBN 3-7643-3567-X
   (hard : acid-free)
       1. Stochastic analysis -- Congresses.  I. Cruzeiro, A.B. (Ana Bela
   Ferreira), 1957-   .  II. Zambrini, Jean-Claude.   III. Lisbon
   Conference on Stochastic Analysis and Applications (1st : 1989)
   IV.  Series:  Progress in probability : 26.
   QA274.2.S7713  1991                              91-22461
   519.2--dc20                                      CIP


Printed on acid-free paper.

© Birkhäuser Boston 1991

ISBN 0-8176-3567-X
ISBN 3-7643-3567-X


Typeset by Authors in TeX and AMS TeX.
Printed and bound by Quinn Woodbine, Woodbine, N.J.
Printed in the U.S.A.

9 8 7 6 5 4 3 2 1

# Contents

# Preface

At the end of the summer 1989, an international conference on stochastic analysis and related topics was held for the first time in Lisbon (Portugal). This meeting was made possible with the help of INIC and JNICT, two organizations devoted to the encouragement of scientific research in Portugal.

The meeting was interdiciplinary since mathematicians and mathematical physicists from around the world were invited to present their recent works involving probability theory, analysis, geometry and physics, a wide area of cross fertilization in recent years.

Portuguese scientific research is expanding fast, these days, faster, sometimes, than the relevant academic structures. The years to come will be determinant for the orientation of those young Portuguese willing to take an active part in the international scientific community.

Lisbon's summer 89 meeting should initiate a new Iberic tradition, attractive both for these researchers to be and, of course, for the selected guests. Judging by the quality of contributions collected here, it is not unrealistic to believe that a tradition of "southern randomness" may well be established.

# Contributors

*S. Albeverio,* Fakultät für Mathematik, Ruhr Universität, D-4630 Bochum, Germany and BiBos Research Center, Universität Bielefeld, Bielefeld, Germany, and SFB 237-Essen, Bochum, Düsseldorf; CERFIM, Locarno, Switzerland

*Lars Andersson,* Department of Mathematics, Royal Institute of Technology, S-10044 Stockholm, Sweden

*Maria Teresa Arede,* Faculdade de Engenharia-DEMEC, R. dos Bragas, P-4000 Porto, Portugal

*Dominique Bakry,* Laboratoire de Statistique et Probabilités, Université Paul Sabatier, 118, route de Narbonne, 31062, Toulouse Cedex, France

*Ana Bela Cruzeiro,* Instituto de Fisica e Matemática (INIC), Av. Prof. Gama Pinto, 2, 1699 Lisboa Codex, Portugal

*K. D. Elworthy,* Mathematics Institute, University of Warwick, Coventry CV4 7AL, England

*Ezra Getzler,* Department of Mathematics, Massachusetts Institute of Technology, Cambridge, MA 02139, USA

*K. Iwata,* Fakultät für Mathematik, Ruhr Universität, D-4630 Bochum, Germany, and SFB-237, Bochum-Essen-Düsseldorf, Germany

*T. Kolsrud,* Department of Mathematics, Royal Institute of Technology, S-10044 Stockholm, Sweden

*Zhi Ming Ma,* BiBos Research Center, D-4800 Bielefeld, Germany, on leave of absence from the Institute of Applied Mathematics, Academia Sinica, Beijing, People's Republic of China

*Paul Malliavin,* 10 rue Saint-Louis en l'Ile, 75004 Paris, France

*David Nualart,* Facultat de Matemàtiques, Universitat de Barcelona, Gran Via, 08007 Barcelona, Spain

*Etienne Pardoux,* Mathématiques, URA 225, Université de Provence, 1331 Marseille Cedex 3, France

*Gunnar Peters,* Department of Mathematics, Royal Institute of Technology, S-10044 Stockholm, Sweden

*Laurent Saloff-Coste,* Université de Paris VI-CNRS, Analyse Complexe et Géométrie (URA D213), 4, place Jussieu - Tour 46 75752, Paris Cedex 05, France

*Daniel W. Stroock,* Massachusetts Institute of Technology, Room 2-272, Cambridge, MA 02139, USA

*Jean Claude Zambrini,* Department of Mathematics, Royal Institute of Technology, S-10044 Stockholm, Sweden, and Instituto de Fisica e Matemática (INIC), Av. Prof. Gama Pinto, 2, 1699 Lisboa Codex, Portugal

# CONFORMALLY INVARIANT RANDOM FIELDS AND PROCESSES—OLD AND NEW

S. Albeverio*#‡, K. Iwata*# , T. Kolsrud**

\* Ruhr-Universität Bochum, FRG
\*\* Kungliga Tekniska Högskolan, Stockholm, Sweden
\# SFB 237, Bochum-Essen-Düsseldorf, FRG
‡BiBo-S Research Centre, Universität Bielefeld, FRG, and
CERFIM, Locarno, Switzerland

**1. Introduction.** In [7], based on earlier articles of which we refer to [1, 2, 3, 6], we studied the equation

$$(1.1) \qquad \bar{\partial} A = F,$$

where $A$ and $F$ are four-component random fields of four variables, and $\bar{\partial}$ is a quaternionic version of the classical Cauchy-Riemann operator in $\mathbf{C}$. Here $F$, the given source, is a white noise, not necessarily gaussian.

It is of interest to consider objects with specific invariance properties, and a large part of [7] is devoted to examining the invariance under the euclidean group $SO(4) \odot \mathbf{R}^4$ (semi-direct product). It turns out that under appropriate conditions on the Lévy measure of $F$, the latter is euclidean invariant in that $a \cdot F =^d F$, where $a \cdot F$ denotes the action of $a \in SO(4) \odot \mathbf{R}^4$ on $F$, and $=^d$ denotes equality in law.

Furthermore, under these conditions on $F$ (1.1) can be solved so that also $A$ becomes euclidean invariant. The meaning of this is that we can find two representations of the euclidean group with $\bar{\partial}$ as intertwining operator.

The basic motivation behind this construction lies in Quantum Field Theory and some steps towards constructions in Minkowski space were taken in [7]. (For further work in QFT, see [2-5].) During the last decade or so, field theories based on conformal invariance have attracted great interest, and it turns out that it is possible to solve (1.1) and have some conformal invariance. The conformal transformations are of the form $x \to (ax + b)(cx + d)^{-1}$, where the variable $x$ and the coefficients $a, b, c, d$ are quaternions.

In the present paper we shall study some examples of conformally invariant random fields given by the equation (1.1), but in dimension two. This has already been studied by Albeverio and Høegh-Krohn (unpublished) and by Osipov [22]. The conformal structure in two dimensions is overwhelmingly rich, which one hopes will lead to new interesting models. Another appealing aspect is the simplification resulting from $\mathbf{C}$, as opposed to the quaternions, being commutative. As a consequence reflection symmetry and positivity hold in two dimensions, as remarked by Osipov. For further interesting results in the quaternionic case, see also Tamura [24].

The general set up is that we have two complex line bundles $E$ and $E'$ over a Riemann surface $M$. $A$ and $F$ are sections of $E$ and $E'$ respectively, and connected by the Cauchy-Riemann operator as in (1.1). These bundles depend in different ways on a real parameter $\kappa$, which enters through weights: Basically, $A$ and $F$ transforms, under the relevant group of Möbius transformations, as generalised meromorphic forms of in general fractional order. These transformations act projectively on $E$ and $E'$.

Chapter 2 is a partial catalogue of known results on conformally invariant random processes, primarily, and fields. The material in 2.2 ([17]) is a mild extension in probabilistic form of results due to Fuglede [11].

In Chapter 3 a heuristic but rather detailed construction of random fields with invariance under $SL(2, \mathbf{R})$ is sketched. It corresponds to the case where $M$ is the complex upper half-plane and $E = M \times \mathbf{C}$. Further examples can be obtained by locally, and w.r.t. the first component, forming the quotient w.r.t. suitable subgroups of $SL(2, \mathbf{R})$. Needless to say, one could similarly treat the cases where $M$ is the Riemann sphere or a torus.

To complete the construction that is sketched in Ch. 3, we require certain invariant measures. The existence of such measures is taken care of in Chapter 4, where we also clarify some relevant aspects of representation theory.

The general picture is far from clear at present. Still we think the examples presented below are sufficiently interesting to motivate and justify an incomplete and to some extent heuristic treatment.

## 2. Examples.

**2.1** The classical result, due to Lévy in the 1940s, see [19, 21], is on the conformal invariance of complex brownian motion $(Z_t, t \geq 0)$. If $f: D \to \mathbf{C}$ (or the Riemann sphere $P^1(\mathbf{C})$) is holomorphic or anti-holomorphic on an open set $D \subset \mathbf{C}$ (or $P^1(\mathbf{C})$), then

$$W_t = f(Z_{t \wedge T}), \quad T = \inf\{t > 0: Z_t \notin D\},$$

is, after a time-shift, brownian motion in $f(D)$. The argument is simple, one just uses Ito's formula to check that $W$ is a conformal martingale, and the conformality of $W$ comes from $f$ being conformal, i.e. $\partial f \cdot \overline{\partial} f = 0$.

**2.2** A far-reaching generalisation of this result to diffusion processes is as follows. Let $M = (M, g)$ and $N = (N, h)$ be riemannian manifolds with metric connections, i.e. $\nabla_M g = 0$ and $\nabla_N h = 0$. In local co-ordinates

$x^1, \ldots x^m$, $^M\Gamma_{ij}^k$ denotes the Christoffel symbols associated with $\nabla_M$. The brownian motion $X$ given by the triplet $(M, g, \nabla_M)$ is the strong solution of the Ito equation

$$dX_t^k = -\tfrac{1}{2}g^{ij}\,{}^M\Gamma_{ij}^k(X_t)dt + a_j^k dB_t^j, \quad 1 \le k \le m,$$

where $\sum_k a_k^i a_k^j = g^{ij}$, the summation convention is employed, and $B$ is a euclidean brownian motion.

We consider now a map $f \colon M \to N$, we let $Z = f(X)$, and we ask when is $Z$ a time-shift of the brownian motion associated with $(N, g, \nabla_N)$?

One can show that this holds if and only if two conditions are fulfilled. The first one, corresponding to the requirement that $W$ is a martingale in the preceeding section 2.1, is that $f$ is a (generalised) harmonic mapping in the sense that the non-linear equations

$$\tfrac{1}{2}\Delta_M f^\alpha + {}^N\Gamma_{\beta\gamma}^\alpha \circ f\langle df^\beta, df^\gamma\rangle = 0, \quad 1 \le \alpha \le n,$$

are satisfied.

The second condition is that $f$ must be (horizontally) conformal in that there is a function $\lambda$ such that

$$\langle df^\alpha, df^\beta\rangle = \lambda^2 h^{\alpha\beta} \circ f, \quad 1 \le \alpha, \beta \le n.$$

In an analytic formulation, and for the Levi-Civita connections, these results were shown by Fuglede in [11]. The connection between martingales in manifolds and harmonic mappings was first observed by Meyer [20] (also in the symmetric (Levi-Civita) case), and later developed by Darling [10]. Probabilistic but non-geometric versions of related results have been developed in a series of papers by Øksendal and Øksendal-Csink, see e.g. [21], [9].

**2.3** The group of Möbius transformations of the unit disk onto itself is isomorphic to $SU(1,1)$. Hence, if we let $D$ be the unit disk and vary $f$ among the Möbius transformations $D \to D$ we see that complex brownian motion on $D$ is $SU(1,1)$-invariant. Similarly for the upper half-plane and $SL(2, \mathbf{R})/\pm I$.

A more general and related example is obtained from brownian motion on $S^{d-1}$, which is invariant under $SO(d, 1)$. This follows from Stroock's construction of $BM(S^{d-1})$ (see [16, 23]). An interesting special case is obtained when $d = 3$, in which case we get invariance of brownian motion on the Riemann sphere under the Lorentz group $SO(3, 1)$, the covering group of which is $SL(2, \mathbf{C})$.

**2.4** Lévy's result has been generalised in various other directions. In the 1960s it was shown by Hida and collaborators, see [12-14], that stable processes are conformally invariant.

We recall that an $\mathbf{R}^d$-valued Markov process $(X_t, t \geq 0)$ is stable of order $0 < p < 2$ when it is translation invariant and

$$\mathbf{E}[e^{i\langle\xi,X_t\rangle}|X_0 = 0] = e^{-c_p t|\xi|^p}.$$

(The limiting case $p = 2$ corresponds to brownian motion.) Corresponding results for random fields in two dimensions are treated in Ch. 4 below.

We mention finally the article Hida-Lee-Lee [15] in which conformal invariance for certain gaussian random fields in $\mathbf{R}^d$ is treated.

**3. Conformally Invariant Random Fields in Two Dimensions.** Consider the group $SL(2, \mathbf{C})$, where $SL(2, \mathbf{C})$ $(SL(2, \mathbf{R}))$ denotes all $2 \times 2$ matrices with complex (real) entries and determinant one. Any element of $SL(2, \mathbf{C})$ gives rise to a (Möbius) transformation of the complex projective line $P^1(\mathbf{C})$ (the Riemann sphere) onto itself by[1]

$$z \to \frac{az + b}{cz + d} = gz, \quad g = \begin{pmatrix} a & b \\ c & d \end{pmatrix} \in SL(2, \mathbf{C}).$$

We denote by $f$ the function

$$f(z; w) = \frac{1}{z - w}.$$

If $g = \begin{pmatrix} a & b \\ c & d \end{pmatrix}$, then $g^{-1} = \begin{pmatrix} d & -b \\ -c & a \end{pmatrix}$, and one finds that

$$f(g^{-1}z; w) = \frac{-c}{cw + d} + \frac{1}{(cw + d)^2} f(z; gw).$$

Consequently, if $A$ is a meromorphic function of the form

$$(3.1) \qquad A(z) = \sum_j \frac{\alpha_j}{i\pi(z - z_j)},$$

with a convergent series, then

$$(3.2) \qquad A(g^{-1}z) = \sum_j \frac{-c\alpha_j}{cz_j + d} + \sum_j \frac{\alpha_j}{(cz_j + d)^2} \frac{1}{i\pi(z - gz_j)}.$$

We assume that the $\alpha_j$ and $z_j$ are such that also the series in (3.2) converge. Suppose now that $A$ is as in (3.1) with all $z_j$ in the upper half-plane

$$\mathbf{C}^+ = \{z : \operatorname{Im} z > 0\}.$$

---

[1] To obtain a one-to-one correspondence we should really consider the group $SL(2,\mathbf{C})/\pm I$ here. Similarly for $SL(2,\mathbf{R})$ and other Möbius groups.

If $g \in G \equiv SL(2, \mathbf{R})$, then neither $-cz+a$ nor $cz+d$ vanish for $z \in \mathbf{C}^+$ and since $\operatorname{Im} z > 0$ it is possible to form any real power of these functions by choosing a logarithm depending on $g$ in general. The powers so obtained are holomorphic functions in $\mathbf{C}^+$. We therefore define, for real $\kappa$ and $g \in G$,

$$(3.3) \qquad g \cdot A(z) \equiv (-cz+a)^{2\kappa-2} A(g^{-1}z).$$

Since the multiplier $(-cz+a)^{2\kappa-2}$ is holomorphic, we obtain

$$\overline{\partial}(g \cdot A)(z) = (-cz+a)^{2\kappa-2} \sum_j \frac{\alpha_j}{(cz_j+d)^2} \delta_{gz_j}$$

$$= \sum_j (-cgz_j+a)^{2\kappa-2} \frac{\alpha_j}{(cz_j+d)^2} \delta_{gz_j}, \quad z \in \mathbf{C}^+,$$

where we used that $f(z)\delta_a = f(a)\delta_a$. Since $g$ has determinant one, it is easily seen that $-cgw + a = 1/(cw+d)$, hence

$$(3.4) \qquad \overline{\partial}(g \cdot A)(z) = \sum_j (cz_j+d)^{-2\kappa} \alpha_j \delta_{gz_j}(z).$$

Suppose now that $F$ is a random measure on $E$ of Poisson type. $F$ may be defined via its characteristic functional or Fourier transform as

$$\mathbf{E}\{e^{-i\langle\beta, F(\Lambda)\rangle}\}$$

$$= \exp\left\{ -\int_\Lambda \int_{\mathbf{C}^\times} \left(1 - e^{-i\langle\beta,\alpha\rangle} + i\frac{\langle\beta,\alpha\rangle}{1+|\alpha|^2}\right) \mu(dxdy, d\alpha)\right\}.$$

Here $\mu \geq 0$ is a Radon measure on $\mathbf{C}^+ \times \mathbf{C}^\times$, where $\mathbf{C}^\times = \mathbf{C}\backslash\{0\}$, satisfying

$$\int_{\mathbf{C}^\times} \frac{|\alpha|^2}{1+|\alpha|^2} \mu(d\alpha, \Lambda) < \infty$$

for any relatively compact Borel set $\Lambda$ in $\mathbf{C}^+$. If the measure $\mu$ is nice, this means that

$$(3.5) \qquad F = \sum_j \alpha_j \delta_{z_j},$$

where the $\alpha_j$ are complex numbers and the points $z_j$ are all in $\mathbf{C}^+$. Furthermore, the pairs $(z_j, \alpha_j)$ are iid random variables, a.s. finitely many, and their number a Poisson distributed random variable. We refer to [3], section 2.6 ff., for further details.

If we now define the random field $A$ as

$$(3.6) \qquad A(z) \equiv \sum_j \frac{\alpha_j}{i\pi(z-z_j)},$$

then clearly

$$(3.7) \qquad \bar{\partial}A = F.$$

The relations (3.2) and (3.3) are still in force. Hence we may conclude from (3.4) and (3.7) that

$$(3.8) \qquad \bar{\partial}(g \cdot A) = g \cdot F,$$

where $g \in G$ acts on the random variables $(z_j, \alpha_j)$ so that

$$(3.9) \qquad (z, \alpha) \rightarrow (gz, (cz + d)^{-2\kappa}\alpha), \quad g = \begin{pmatrix} a & b \\ c & d \end{pmatrix} \in G,$$

i.e.

$$g \cdot F \equiv \sum_j (cz_j + d)^{-2\kappa} \alpha_j \delta_{gz_j}.$$

It is clear from this that if $\mu$ is invariant under the transformation (3.9), then

$$F \overset{d}{=} g \cdot F, \quad g \in G,$$

i.e. $F$ is conformally invariant, and this goes for $A$ too. It should be said here that if $\mu$ is too wild for (3.5) to hold, then the argument given here must be adjusted.

   In the next chapter we shall prove that there are lots of invariant measures.

## 4. Representation theory. Existence of invariant measures.
The aim of this chapter is to somewhat remedy the lack of precision in the preceeding one. We start by considering the action of $SL(2, \mathbf{R})$ proposed above.

   $G = SL(2, \mathbf{R})$ acts on $E = \mathbf{C}^+ \times \mathbf{C}$ in the following way. For each $\kappa \in \mathbf{R}$ we let

$$J^\kappa(z, g) = (cz + d)^{-2\kappa}, \quad g = \begin{pmatrix} a & b \\ c & d \end{pmatrix} \in G, \quad z \in \mathbf{C}^+.$$

This means that if $\kappa \notin \frac{1}{2}\mathbf{Z}$ we have to choose a logarithm depending on $g$ to define $J^\kappa(z, g)$. In general we have

$$(4.1) \qquad J^\kappa(z, g_1 g_2) = J^\kappa(z, g_1) J^\kappa(g_1 z, g_2) e^{4\pi i k \kappa}$$

for some $k = k(g_1, g_2) \in \mathbf{Z}$, which reduces to

$$(4.2) \qquad J^\kappa(z, g_1 g_2) = J^\kappa(z, g_1) J^\kappa(g_1 z, g_2)$$

when $\kappa$ is a half-integer.

We define

(4.3)
$$g \cdot (z, \alpha) = (gz, J^{\kappa}(z, g)\alpha).$$

Then (4.1) implies that this is in general only a projective action of $G$ on $E$ in that
$$(g_1 g_2) \cdot (z, \alpha) = g_1 \cdot (g_2 \cdot (z, e^{i\theta}\alpha))$$
for some real number $\theta$. When $\kappa$ is a half-integer, this is an honest action: $(g_1 g_2) \cdot (z, \alpha) = g_1 \cdot (g_2 \cdot (z, \alpha))$, as we see from (4.2).

We summarise in the following

**(4.4) Proposition.** *Eq. (4.3) defines a projective action of $G = SL(2, \mathbf{R})$ on $E = \mathbf{C}^+ \times \mathbf{C}$ for any real number $\kappa$. When $\kappa$ is a half-integer this is a proper group action.*

We now turn to the existence of invariant measures. It is convenient to denote $\mathbf{C}^+ \times \mathbf{C}^\times$ by $E_0$. We consider measures on $E_0$ on the form
$$\mu(dxdy, d\alpha) = \frac{dxdy}{y^{2-q}} \frac{d\alpha}{|\alpha|^{2+p}}.$$

The Poincaré area $dxdy/y^2$ is invariant under $G$, and $d\alpha/|\alpha|^2$ is dilation invariant. Hence $\mu$ transforms as
$$\mu(dxdy, d\alpha) \longrightarrow \mu(dxdy, d\alpha) \cdot |cz + d|^{2p\kappa - 2q},$$
so $\mu$ is invariant if $q = p\kappa$. We also have to take the integrability condition on $\mu$ into account. This leads to $0 < p < 2$ and corresponds to stable random fields. (Cf. section 2.4.)

It follows from this that any measure on the form

(4.5)
$$\frac{dxdy}{y^2} \frac{d\alpha}{|\alpha|^2} \int \frac{y^{p\kappa}}{|\alpha|^p} \lambda(dp),$$

where $\lambda \geq 0$ is a $\sigma$-finite measure, not necessarily on the interval $(0,2)$, is invariant.

More generally we have the following result:

**(4.6) Theorem.** *Let $\nu \geq 0$ be a rotation invariant measure on $\mathbf{C}^\times$ and define a measure $\mu$ on $E_0 = \mathbf{C}^+ \times \mathbf{C}^\times$ by*

$$\mu(dxdy, d\alpha) \equiv \frac{dxdy}{y^2} \nu(y^{-\kappa} d\alpha).$$

*Then, for any real number $\kappa$, $\mu$ is invariant under the action (4.3) of $G = SL(2, \mathbf{R})$.*

*Proof.* The definition of $\mu$ means that
$$\int_{E_0} f(z, \alpha) \mu(dxdy, d\alpha) = \int_{E_0} f(z, y^{\kappa}\alpha) \frac{dxdy}{y^2} \nu(d\alpha).$$

Hence

$$\int_{E_0} f(g \cdot (z, \alpha)) \, \mu(dxdy, d\alpha)$$

$$= \int_{E_0} f(gz, (y/|cz + d|^2)^\kappa (cz + d)^{2\kappa} \alpha) \, \frac{dxdy}{y^2} \nu(d\alpha)$$

$$= \int_{E_0} f(gz, y^\kappa e^{i\theta} \alpha) \frac{dxdy}{y^2} \nu(d\alpha) = \int_{E_0} f(z, \alpha) \, g^{-1} \left( \frac{dxdy}{y^2} \right) \nu(y^{-\kappa} e^{-i\theta} d\alpha)$$

$$= \int_{E_0} f(z, \alpha) \, \mu(dxdy, d\alpha).$$

*(4.7) Remarks.* This type of construction can of course be carried out in various other situations. We may put restrictions on $\kappa$, and we can replace the upper half-plane $\mathbf{C}^+$ by some other hyperbolic Riemann surface $M = \mathbf{C}^+/\Gamma$, where $\Gamma$ is a discontinuous subgroup of $G$. If $\kappa$ is a rational number it suffices to have $\nu$ invariant under finitely many rotations, i.e. under some $\mathbf{Z}_n$.

If $E$ is a bundle over $M = \mathbf{C}^+/\Gamma$, i.e. if $E$ locally looks like $M \times \mathbf{C}$, we obtain an invariant measure for the action of $\Gamma$ simply by restricting $\mu$ in the first variable to a fundamental domain for $\Gamma$. It is known that for many subgroups $\Gamma$ of $G$ the action (4.3) can be redefined to an ordinary one. For instance, this is the case for the modular group $SL(2, \mathbf{Z})$, and therefore also for all of its subgroups. See e.g. Mackey [18], Ch. 30. Finally, if we replace the group considered by its universal covering group, we obtain a true action for any choice of $\kappa$.

**(4.8)** It is well known that given a group action on a (nice) topological space and a measure which is invariant under this action, we obtain a unitary representation of the group in question. (See e.g. Barut-Raczka [8], Ex. 2, Ch. V.1.) Here we have a similar situation, but–as we have already seen– the action is in general only projective. One can however make a slight modification to obtain unitary representations also in the case considered here, as we shall explain now.

We consider functions $\phi$ on $E$ which are rotation invariant in $\alpha$, which means that for any real number $\theta$

(4.9) $$\phi(z, \alpha) = \phi(z, e^{i\theta} \alpha).$$

For such $\phi$s and for any real $\kappa$ we have

$$(g_1 g_2) \cdot \phi = g_1 \cdot (g_2 \cdot \phi),$$

where

(4.10) $$(g \cdot \phi)(z, \alpha) \equiv \phi(g^{-1}(z, \alpha)) = \phi(g^{-1}z, J^\kappa(z, g^{-1}z)\alpha)), \ g \in G.$$

Let now $\mu$ be as in Theorem 4.6, and let $H$ be the closed subspace of $L^2(E, d\mu)$ consisting of functions that are rotation invariant in the second variable. One checks easily that (4.10) gives a unitary representation of $G$ on the Hilbert space $H$.

The remarks in (4.7) apply here too. Condition (4.9) can be weakened depending on $\kappa$. Similarly, we may consider certain subgroups $\Gamma$ of $G$ which leads to induced representations of $G$.

*Acknowledgements.* We are grateful for support from the Swedish Natural Science Research Council, NFR, the Swedish Board of Technical Development, STUF, the Royal Swedish Academy of Sciences, and the German Sonderforschungsbereich, project SFB 237, "Unordnung und große Fluktuationen", Essen/Bochum/Düsseldorf. Many thanks also to the Mathematical Departments at Ruhr-Universität Bochum, and the Royal Institute of Technology, Stockholm. Many thanks to Ana Bela Cruzeiro, who enabled us to participate in an interesting conference in a nice part of the world.

*References.*
[1] Albeverio, S. and Høegh-Krohn, R.: *Construction of interacting local relativistic quantum fields in four space-time dimensions*, Phys. Lett. **B200**, 108-114 (1988). Erratum in ibid. **202** p. 621 (1988).

[2] Albeverio, S., Høegh-Krohn, R., and Holden, H.: *Markov cosurfaces and gauge fields*. Acta Phys. Austr. Suppl. XXVI, 211-31 (1984).

[3] Albeverio, S., Høegh-Krohn, R., Holden, H. and Kolsrud, T.: *Representation and construction of multiplicative noise*. J. Funct. Anal. **87** (1989), 250-272.

[4] Albeverio, S., Høegh-Krohn, R., Holden, H. and Kolsrud, T.: *A covariant Feynman-Kac formula for unitary bundles over Euclidean space*. In G. DaPrato and L. Tubaro (Eds), Proc. Conf. Stochastic Partial Differential Equations II, Trento Feb. 1988. Springer Lect. Notes in Math.**1390**, 1989.

[5] Albeverio, S., Høegh-Krohn, R., Holden, H. and Kolsrud, T.: *Construction of quantised Higgs-like fields in two dimensions*. Phys. Lett. B. **222** (1989), 263-268.

[6] Albeverio, S., Høegh-Krohn, R. and Iwata, K.: *Covariant markovian random fields in four space-time dimensions with nonlinear electromagnetic interaction*. In 'Applications of self-adjoint extensions in quantum physics.' Proc. Dubna Conference 1987, Editors P. Exner, P. Seba, Springer Lect. Notes in Phys. **324**, 1989.

[7] Albeverio, S., Iwata., K. and Kolsrud., T., *Random Fields as Solutions of the Inhomogeneous Quaternionic Cauchy-Riemann Equation. I. Invariance and Analytic Continuation*. Preprint Stockholm/BiBoS 1989. Commun. Math. Phys. **132**, 555-580 (1990).

[8] Barut, A.O. and Raczka, R.: *Theory of group representations and applications*, 2nd ed. World Scientific, Singapore 1986.

[9] Csink, L. and Øksendal, B. *Stochastic harmonic morphisms: functions mapping the paths of one diffusion into the paths of another.* Ann. Inst. Fourier (Grenoble) **33** (1983), 219-240.

[10] Darling, R.W.R. *Martingales in manifolds–Definitions, examples, and behaviour under maps.* In: "Séminaire de Probabilités XVI, 1980/81. Supplément: Géométrie Différentielle Stochastique" (Eds J. Azéma and M. Yor), 217-236. Lect. Notes in Math. 921, Springer (1982).

[11] Fuglede, B. *Harmonic morphisms between Riemannian manifolds.* Ann. Inst. Fourier (Grenoble) **28** (1978), 107-144.

[12] Hida, T., *Sur l'invariance projective pour les processus symétriques stables.* C. R. Acad. Sc. Paris t. **267** (1968), 821-823.

[13] Hida, T., Kubo, I., Nomoto, H. and Yosizawa. H., *On projective invariance of Brownian motion.* Pub. Res. Inst. Math. Sci. Kyoto Univ. **4** (1969), 595-609.

[14] Hida, T., Stationary stochastic processes. Princeton Lecture Notes Series in Math., 1970.

[15] Hida, T., Lee, K.-S. and Lee, S.-S., *Conformal invariance of white noise.* Nagoya Math. J. **98** (1985), 87-98.

[16] Ikeda, N. and Watanabe, S. Stochastic differential equations and diffusion processes. North-Holland/Kodansha, 1981.

[17] Kolsrud, T. *On maps preserving geometrical semimartingale diffusions.* Manuscript 1986/87/88/89/?

[18] Mackey, G.W., Unitary group representations in physics, probability, and number theory. Mathematical Lecture Notes Series, Benjamin, 1978.

[19] McKean, H.P. Stochastic integrals. Academic Press, 1969.

[20] Meyer, P.-A. *A differential geometric formalism for the Itô calculus.* In: "Stochastic Integrals, Proceedings, LMS Durham Symposium, 1980" (Ed D. Williams), 256-270. Lect. Notes in Math. 851, Springer, 1981.

[21] Øksendal, B. *Finely harmonic morphisms, Brownian path preserving functions and conformal martingales.* Inventiones Math. **75** (1984), 179-187.

[22] Osipov, E. *Two-dimensional random fields as solutions of stochastic differential equations.* Bochum preprint 1989.

[23] Stroock, D., *On the growth of certain stochastic integrals.* Z. Warsch. verw. Geb., **18** (1971), 340-344.

[24] Tamura, H. *Nonlinear electromagnetic fields confine charges.* Kanazawa preprint 1989.

March 1990

S. Albeverio & K. Iwata        T. Kolsrud
Dept. of Math.                 Dept. of Math.
University of Bochum           Royal Institute of Technology
D-4630, Bochum                 S-100 44 Stockholm
FRG                            Sweden

# Diffusion Processes with
# singular Dirichlet forms

by

Sergio Albeverio[*,**,#], Zhi Ming Ma[**,***]

## ABSTRACT

In this paper we construct diffusion processes associated with such Dirichlet forms which are so singular that their form domains contain no continuous functions different from the zero function. In quantum theory such singular Dirichlet forms correspond to Schrödinger operators with potentials which are singular on each neighbourhood of every point.

* Fakultät für Mathematik, Ruhr–Universität,
  D-4630 BOCHUM (FRG)
** BiBoS Research Center, D-4800 Bielefeld (FRG)
*** On leave of absence from Inst. of Appl. Mathematics, Academia Sinica, Beijing
 # SFB 237 – Essen, Bochum, Düsseldorf;
   CERFIM, Locarno

**Keywords:** Dirichlet form, $\mathcal{E}$-nest, perfect process, diffusion process, nowhere Radon smooth measure, perturbed Dirichlet form, singular Schrödinger operator

## 1. Introduction

We first recall the definition of Dirichlet forms. Consider a metrizable topological space $X$ and a positive $\sigma$-finite Borel measure $m$ on $X$. One calls $(\mathcal{E}, \mathcal{F})$ a nonnegative quadratic form on $L^2(X, m)$ if $\mathcal{F}$ is a linear dense subspace of $L^2(X, m)$ and $\mathcal{E} : \mathcal{F} \times \mathcal{F} \to R$ is a nonnegative symmetric bilinear form. $(\mathcal{E}, \mathcal{F})$ is said to be closed if $\mathcal{F}$ is complete with respect to the metric given by the inner product $\mathcal{E}_1(f, g) := \mathcal{E}(f, g) + (f, g)$, $\forall f, g \in \mathcal{F}$. A closed nonnegative quadratic form is called a <u>Dirichlet form</u> if $u \in \mathcal{F}$ implies $(0 \vee u) \wedge 1 := u^* \in \mathcal{F}$ and $\mathcal{E}(u^*, u^*) \leq \mathcal{E}(u, u)$. The theory of Dirichlet forms is a powerful combination of functional analysis and stochastic analysis for a general potential theory associated with $L^2$-spaces. In this connection see the fundamental works of Fukushima [F2] and Silverstein [S]. There is a one to one correspondence between the symmetric (strongly continuous) Markov semigroups and the Dirichlet forms on $L^2(X, m)$. In fact, if $A$ is the generator of a symmetric Markov semigroup $(T_t)_{t>0}$, then the quadratic form $(\mathcal{E}, \mathcal{F})$ given by $\mathcal{E}(u, v) = (\sqrt{-A}u, \sqrt{-A}v)$, with $\mathcal{F} = D(\sqrt{-A})$ ($D(\sqrt{-A})$ denotes the domain of $\sqrt{-A}$) is a Dirichlet form. Viceversa, to any Dirichlet form $(\mathcal{E}, \mathcal{F})$ there exists a unique negative definite self-adjoint operator $A$ on $L^2(X, m)$, such that $\mathcal{F} = D(\sqrt{-A})$ and $\mathcal{E}(u, v) = (\sqrt{-A}u, \sqrt{-A}v)$, and such that the semigroup $T_t = e^{tA}$, $t > 0$, is a symmetric Markov semigroup on $L^2(X, m)$. A Markov process $(X_t)_{t \geq 0}$ with state space $X$ and transition function $(P_t)_{t \geq 0}$ is said to be associated with a Dirichlet form $(\mathcal{E}, \mathcal{F})$ if $T_t u = P_t u$ $m$ a.e. for each $u \in L^2(X, m)$ and $t > 0$, where $(T_t)_{t>0}$ is the unique Markov semigroup corresponding to $(\mathcal{E}, \mathcal{F})$ in the above manner. Of course, in order to have a reasonable Markov process associated with $(\mathcal{E}, \mathcal{F})$, some conditions on $(\mathcal{E}, \mathcal{F})$ (as well as on $X$) must be imposed. In the case where the basic underlying space $X$ is locally compact, a well known condition is that $\mathcal{F} \cap C_0(X)$ ($C_0(X)$ denotes all the continuous functions with compact supports) is dense both in $\mathcal{F}$ (with $\mathcal{E}_1$-metric) and in $C_0(X)$ (with supremum norm). One calls such a Dirichlet form <u>regular</u>. M.Fukushima and M.L. Silverstein first constructed a Hunt process associated with a regular Dirichlet form ([F1], [F2]). There have also been results concerning the existence of a diffusion process associated with a Dirichlet form where $X$ is not necessarily locally compact, see [AH1]-[AH3], [AR1,2], [F3] and [K] (and references therein).

A common condition, imposed by all the above mentioned works concerning the existence of a reasonable Markov process associated with a given Dirichlet form $(\mathcal{E}, \mathcal{F})$, is that:

$$\mathcal{F} \cap C(X) \text{ is } \mathcal{E}_1 - \text{dense in } \mathcal{F}, \qquad (1.1)$$

where $C(X)$ is the family of all continuous functions on X. One might think of the conjecture that the condition (1.1) would be a necessary condition for the existence of a resonable Markov process, at least for the existence of a diffusion process, associated with $(\mathcal{E}, \mathcal{F})$.

Based on some general results of our recent work, we shall show in this paper that the above conjecture is not true. In fact, we shall obtain diffusion processes associated with the Dirichlet forms which are so singular that the form domain does not contain any continuous functions other than the zero function. In quantum theory such singular Dirichlet forms are associated with Schrödinger operators which are singular on each neighbourhood of every point.

Let us now describe shortly the organization of this paper. In Section 2 we first recall the concept of smooth measures in terms of the capacity relative to a given regular Dirichlet form $(\mathcal{E}, \mathcal{F})$, and then show that there always exist smooth measures $\mu$ which are nowhere Radon in the sense that $\mu(G) = \infty$ for all non-empty open set $G \subset X$, provided each single-point of $X$ is of zero capacity. In section 3, we prove that the perturbed form $(\mathcal{E}^\mu, \mathcal{F}^\mu)$ with $\mathcal{F}^\mu = \mathcal{F} \cap L^2(X, m)$ and $\mathcal{E}^\mu(f, g) = \mathcal{E}(f, g) + \int f g \mu(dx)$ is again a Dirichlet form provided $\mu$ is smooth. In particular, if $\mu$ is a nowhere Radon smooth measure then $(\mathcal{E}^\mu, \mathcal{F}^\mu)$ is a singular Dirichlet form such that the only element in $\mathcal{F}^\mu \cap C(X)$ is the function identically taking value zero. In section 4 we briefly describe the necessary and sufficient conditions for a Dirichlet form to have associated with it an $m$-perfect process (i.e., a normal strong Markov process with cadlag paths up to life time and with regular resolvent functions), and then we point out those conditions (for having an $m$-perfect process) which together with the local property of the Dirichlet form are the necessary and sufficient conditions for having an associated diffusion process. The proof of these results will appear separately in our forthcoming papers [AM5] and [AM6]. In section 5 we show that the perturbed form $(\mathcal{E}^\mu, \mathcal{F}^\mu)$ always satisfies the necessary and sufficient conditions described in section 4, hence there always exists an $m$-perfect process associated with $(\mathcal{E}^\mu, \mathcal{F}^\mu)$. Moreover, if $(\mathcal{E}, \mathcal{F})$ possesses the local property, then so does $(\mathcal{E}^\mu, \mathcal{F}^\mu)$, and consequently there exists a diffusion process associated with $(\mathcal{E}^\mu, \mathcal{F}^\mu)$. In the final section we apply the above results to the classical case of $(\mathcal{E}, \mathcal{F})$ being the Dirichlet form associated with the Laplacian $\frac{-\Delta}{2}$ on $\mathbb{R}^d$, $d \geq 2$. We prove that in this case the self adjoint operator $H^\mu$ associated with $(\mathcal{E}^\mu, \mathcal{F}^\mu)$ is nothing but the Schrödinger operator $(\frac{-\Delta}{2} + \mu)$. In this case there always exists nowhere Radon smooth measures. For such nowhere Radon smooth measures $\mu$ the form domain $\mathcal{F}^\mu$ of $\mathcal{E}^\mu$ contains no continuous functions other than the zero function and the corresponding Schrödinger operator $(\frac{-\Delta}{2} + \mu)$ is singular on each neighbourhood of every point $x \in \mathbb{R}^d$. Our results show that even in this case there still exists a diffusion process associated with $(\mathcal{E}^\mu, \mathcal{F}^\mu)$, i.e., there is a diffusion process whose generator is $(\frac{\Delta}{2} - \mu)$.

## 2. Nowhere Radon smooth measures

Consider a regular Dirichlet form $(\mathcal{E}, \mathcal{F})$ on $L^2(X, m)$, where $X$ is a locally compact metrizable space and $m$ an everywhere dense Radon measure on $X$.

For $\alpha > 0$, the $\underline{\mathcal{E}_\alpha\text{-norm}}$ on $\mathcal{F}$ is defined by

$$\mathcal{E}_\alpha(f, f) = \mathcal{E}(f, f) + \alpha(f, f), \quad \forall f \in \mathcal{F}, \tag{2.1}$$

where $(\cdot, \cdot)$ denotes the usual inner product in $L^2(X, m)$.

The capacity Cap is defined by

$$\mathrm{Cap}(G) = \inf \{\mathcal{E}_1(u, u) : u \in \mathcal{F}, \, u \geq 1 \quad \text{m.a.e. on } G\} \tag{2.2}$$

for an open set $G$ and

$$\mathrm{Cap}(A) = \inf \{\mathrm{Cap}(G) : \, G \supset A, \, G \quad \text{open}\} \tag{2.3}$$

for an arbitrary set $A$. A Borel measure $\mu$ on $X$ is said to be $\underline{\text{smooth}}$ ($\mu \in S$ in notation) if $\mu$ charges no set of zero capacity and there exists an increasing sequence of compact sets $\{F_n\}$ such that

(i) $\mu(F_n) < \infty$, $\forall n \geq 1$, $\hspace{5cm}$ (2.4)
(ii) $\mu(X_- \cup F_n) = 0$, $\hspace{6.2cm}$ (2.5)
(iii) $\lim_{n \to \infty} \mathrm{Cap}(K - F_n) = 0$, $\quad \forall$compact set $K$. $\hspace{2.5cm}$ (2.6)

We shall say that a Borel measure $\mu$ on $X$ is $\underline{\text{nowhere Radon}}$ if $\mu(G) = \infty$ for all non-empty open set $G$ of $X$.

The following fact was discovered in [AM1] and will play an important role in constructing Dirichlet forms with domain containing no non-zero continuous functions.

**2.1 Theorem** [AM1] Suppose that each single-point set of $X$ is a set of zero capacity. Then, for each smooth measure $\nu$ with $\mathrm{supp}[\nu] = X$, there exists a smooth measure $\mu$ such that $\mu$ is equivalent to $\nu$ and $\mu$ is nowhere Radon.

**Proof** Put $B = \{x \in X : \nu$ is finite on a neighbourhood of $x\}$. If $B = \phi$, then $\nu$ itself is a desired nowhere Radon smooth measure. We assume now $B \neq \phi$. Let $\{x_j\}_{j \geq 1}$ be a countable dense subset of $B$. We choose for each $j$ a decreasing sequence of open sets $\{G_{j,k}\}_{k \geq 1}$ such that

$$\cap_{k \geq 1} G_{j,k} = x_j$$

and

$$\mathrm{Cap}\,(G_{j,k}) + \nu(G_{j,k}) \leq 2^{-k}$$

Such a sequence exists because we have $\mathrm{Cap}(\{x_j\}) = 0$ which implies $\nu(\{x_j\}) = 0$, and $\nu$ is finite on a neighborhood of $x_j$.

We now define for each $j$

$$f_j(x) = \begin{cases} \frac{k}{\nu(G_{j,k})} & \text{when } x \in G_{j,k} - G_{j,k+1}, \ k = 1,2,... \\ 1 & \text{otherwise} \end{cases}$$

Then $f_j$ is bounded outside $G_{j,k}$ for each $k$. In particular, we can choose a positive number $c_j$ such that

$$c_j \sup_{x \in X - G_{j,j}} f_j(x) \le 2^{-j}$$

We define

$$f(x) = \sum_{j \ge 1} c_j f_j(x)$$

and

$$\mu(dx) = f(x)\nu(dx).$$

We claim that $\mu$ is a smooth measure with the required properties. Indeed, since $0 < f < \infty$, it is evident that $\mu = f \cdot \nu$ is equivalent to $\nu$, and consequently $\mu$ charges no set of zero capacity. Let $\{E_n\}_{n \ge 1}$ be an increasing sequence of compact sets satisfying (2.4)–(2.6) with respect to $\nu$. We set for each $n$:

$$G_n = (\cup_{1 \le j \le n} G_{j,2n}) \cup (\cup_{j > n} G_{j,j})$$

and

$$F_n = E_n - G_n.$$

Then $\{F_n\}_{n \ge 1}$ is an increasing sequence of compact sets and it is easy to check that (2.4)–(2.6) hold for $\mu$ and $\{F_n\}_{n \ge 1}$. Thus $\mu$ is a smooth measure. Let $G$ be a non-empty open set of $X$. If $G - B \ne \phi$, then $\mu(G) \ge c\nu(G) = \infty$ by the definition of $B$. Here $c := \sum_{j \ge 1} c_j > 0$. Suppose now $G \cap B \ne \phi$. Then $x_j \in G$ for some $j \ge 1$, and for $k$ large enough we have $G_{j,k} \subset G$. In this case we have

$$\mu(G) \ge c_j \int_{G_{j,k}} \frac{k}{\nu(G_{j,k})}\, \nu(dx) \ge c_j k.$$

Letting $k \to \infty$ we obtain $\mu(G) = \infty$. Thus $\mu$ is as desired. ∎

We remark that the reference measure $m$ is always a smooth measure with $\text{supp}[m] = X$. Hence there always exists a nowhere Radon smooth measure provided each single-point set is a set of zero capacity. The latter propery is satisfied, for example, if $(\mathcal{E}, \mathcal{F})$ is the classical Dirichlet form associated with the Laplacian operator in $\mathbb{R}^d$ with $d \ge 2$. See Section 6 for details.

In the remainder of this section we recall some concepts relative to the capacity which will be used in the remaining sections.

A property is said to hold quasi-everywhere, abbreviated q.e., if it holds except on a set of zero capacity. Two functions defined q.e. on $X$ are said to be quasi-equivalent if they are equal q.e.. An increasing sequence of closed sets $\{F_k\}_{k\geq 1}$ is called a nest if $Cap(X - F_k) \downarrow 0$. A function $u$ defined q.e. on $X$ is said to be quasicontinuous, if there exists a nest $\{F_k\}$ such that $u_{|F_k}$ is continuous for each $k$. Here $u_{|F_k}$ denotes the restriction of $u$ to $F_k$. It is known [Fu2] that for each $u \in \mathcal{F}$, there exists a quasicontinuous function, which we shall always denote by $\tilde{u}$, such that $\tilde{u} = u$ m. a.e.. $\tilde{u}$ is called a quasicontinuous modification of $u$.

## 3. Perturbed Dirichlet forms

In this section we study the perturbation of Dirichlet forms by means of positive continuous additive functionals associated with smooth measures.

Let $(\mathcal{E}, \mathcal{F})$ be as in the previous section. It is well known that there exists a Hunt process $(X_t) := (\Omega, M, M_t, X_t, P_x)$ associated with $(\mathcal{E}, \mathcal{F})$. The cemetery $\Delta$ of $(X_t)$ is adjoint to the state space $X$ in the manner that $X_\Delta := X \cup \{\Delta\}$ is a one point compactification of $X$. Any function $f$ defined on $X$ is automatically extended to $X_\Delta$ by setting $f(\Delta) = 0$. $\zeta$ will always denote the life time of $(X_t)$.

In the setting of Dirichlet spaces, a nonnegative function $(A_t)$ defined on $[0, \infty) \times \Omega$ is said to be a PCAF (positive continuous additive functional) of $(X_t)$ if:
(i) $A_t(\cdot)$ is $\mathcal{F}_t$–measurable, where $\mathcal{F}_t$ is the smallest complete $\sigma$–algebra which contains $\sigma\{X_s : s \leq t\}$;
(ii) there exists a defining set $\Lambda \in \mathcal{F}_\infty$ and an exceptional set $N \subset X$ with $Cap(N) = 0$ such that $P_x(\Lambda) = 1$ for all $x \in X - N$, $\Theta_t\Lambda \subset \Lambda$ for all $t > 0$ ($\Theta_t$ denotes the shift operator on $\Omega$) and for each $w \in \Lambda$, $A.(w)$ is a nonnegative continuous function satisfying: $A_0(w) = 0$, $A_t(w) < \infty$ for $t < \zeta(w)$ and $A_{t+s}(w) = A_t(w) + A_s(\Theta_t w)$ for $s, t \geq 0$.

Two PCAF $(A_t)$ and $(B_t)$ are said to be equivalent if they have a common defining set $\Lambda$ and a common exceptional set $N$ such that $A_t(w) = B_t(w)$ for all $w \in \Lambda$ and $t \geq 0$.

By [F2] Th 5.1.5, for each smooth measure $\mu$, there exists a unique (up to equivalence) PCAF $(A_t^\mu)$ such that

$$\lim_{t \downarrow 0} \frac{1}{t} E_{h \cdot m} \left[ \int_0^t f(X_s) dA_s^\mu \right] = < f \cdot \mu, k > := \int_X h(x)(f \cdot \mu)(dx) \qquad (3.1)$$

for any $\gamma$–excessive function $h(\gamma \geq 0)$ and $f \in \mathcal{B}^+$ ($\mathcal{B}^+$ denotes all non-negative Borel functions on $X$); and the formula (3.1), which is called Revuz formula, specifies a one to one correspondence between the family of all smooth measures and the family of all equivalent classes of PCAF's.

Let $\mu \in S$ correspond to $(A_t^\mu)$. We introduce the notations

$$P_t^\mu f(x) = E_x \left[ e^{-A_t^\mu} f(X_t) \right] \quad \text{for } t > 0 \qquad (3.2)$$

and

$$U_\alpha^\mu f(x) = E_x \left[ \int_0^\infty e^{-\alpha t - A_t^\mu} f(X_t) dt \right] \quad \text{for } \alpha > 0 \qquad (3.3)$$

provided the integrals make sense. The notations $(P_t)_{t>0}$ and $(U_\alpha)_{\alpha>0}$ are understood by taking $\mu \equiv 0$ in (3.2) and (3.3) respectively.

We now consider $(\mathcal{E}^\mu, \mathcal{F}^\mu)$, the perturbation of $(\mathcal{E}, \mathcal{F})$ by a smooth measure $\mu$, which is defined by

$$\mathcal{F}^\mu = \mathcal{F} \cap L^2(X, \mu) \tag{3.4}$$

$$\mathcal{E}^\mu(f, g) = \mathcal{E}(f, g) + <f, g>_\mu \quad , \forall f, g \in \mathcal{F}^\mu, \tag{3.5}$$

where $<f, g>_\mu := \int_X fg\,\mu(dx)$. For $\alpha > 0$ we put also

$$\mathcal{E}^\mu_\alpha(f, g) = \mathcal{E}_\alpha(f, g) + <f, g>_\mu. \tag{3.6}$$

In what follows we fix a smooth measure $\mu$.

**3.1 Lemma**     Let $f \in L^2(X, m)$ and $\alpha > 0$. Then $U^\mu_\alpha f \in \mathcal{F}^\mu$ and

$$\mathcal{E}^\mu_\alpha(U^\mu_\alpha f, g) = (f, g) \quad , \forall g \in \mathcal{F}^\mu \tag{3.7}$$

Moreover, $U^\mu_\alpha f$ is quasicontinuous.

**Proof**     Without loss of generality we may assume $f \geq 0$. Set

$$\phi(x) = E_x \left[ \int_0^\infty e^{-\alpha t}(U^\mu_\alpha f)(X_t)\,dA^\mu_t \right]. \tag{3.8}$$

By the additive property of $(A^\mu_t)$ and the Markov property of $(X_t)$, and applying Fubini's theorem we have

$$U^\mu_\alpha f + \phi = U_\alpha f \tag{3.9}$$

Since $\phi$ and $U_\alpha f$ are all $\alpha$-excessive functions, by [F2] Lemma 3.3.2 we see that $U_\alpha f, \phi \in \mathcal{F}$ and accordingly $U^\mu_\alpha f \in \mathcal{F}$. Moreover, by [F2] Th. 4.3.2 $U_\alpha f$ and $\phi$ are quasicontinuous and consequently so is $U^\mu_\alpha f$. By (3.1) and [F2] Th. 5.3.1, for any $g \in \mathcal{F}$ such that $g$ is a $\gamma$-excessive function for some $\gamma \geq 0$, we have

$$<U^\mu_\alpha f, g>_\mu = \lim_{t \downarrow 0} \frac{1}{t} E_{g \cdot m} \left[ \int_0^\infty e^{-\alpha s}(U^\mu_\alpha f)(X_s)\,dA^\mu_s \right] = \mathcal{E}_\alpha(\phi, g) \tag{3.10}$$

In particular, taking $g = U_\alpha f$, we obtain

$$<U^\mu_\alpha f, U^\mu_\alpha f>_\mu \leq <U^\mu_\alpha f, U_\alpha f>_\mu = \mathcal{E}_\alpha(\phi, U_\alpha f) = (f, \phi) < \infty.$$

Thus $U^\mu_\alpha f \in \mathcal{F} \cap L^2(X, \mu) := \mathcal{F}^\mu$. From (3.9) and (3.10) we see that if $g \in \mathcal{F}$ is expressed by the difference of two $\gamma$-excessive functions in $\mathcal{F}$, then

$$\mathcal{E}_\alpha(U^\mu_\alpha f, g) = (f, g) - <U^\mu_\alpha f, g>_\mu \tag{3.11}$$

Applying [F2] Lemma 1.3.3 and the dominated convergence theorem, it is verified that (3.11) is also valid for each bounded function $g \in \mathcal{F}$ provided $\mu$ is finite and $f$ is bounded. Let $\{F_n\}_{n \geq 1}$ be an increasing sequence of compact sets satisfying (2.4)–(2.6). We set $\mu_n = I_{F_n} \cdot \mu$, $f_n = f \wedge n$, and $g_n = g \wedge n$. For a nonnegative $g \in \mathcal{F}^\mu$, (3.11) yields

$$\mathcal{E}_\alpha(U_\alpha^\mu f_n, g_n) = (f_n, g_n) - < U_\alpha^\mu f_n, g_n I_{F_n} >_\mu \tag{3.12}$$

Again from (3.11) we can prove

$$\lim_{n \to \infty} \mathcal{E}_\alpha^\mu(U_\alpha^\mu f - U_\alpha^{\mu_n} f_n, U_\alpha^\mu f - U_\alpha^{\mu_n} f_n) = 0 \tag{3.13}$$

Now the assertions of the Lemma follow from (3.12), (3.13), [F2] Th.3.14 and the monotone convergence theorem.

**3.2 Theorem**     Let $P_t^\mu$ be defined by (3.2). Then $(P_t^\mu)_{t>0}$ is a $m$-symmetric strongly continuous Markovian semigroup on $L^2(X, m)$, and $(\mathcal{E}^\mu, \mathcal{F}^\mu)$ is the corresponding Dirichlet form associated with $(P_t^\mu)_{t>0}$.

**Proof**     Obviously $(P_t^\mu)_{t>0}$ is a Markovian semigroup on $L^2(X, m)$ with $(U_\alpha^\mu)_{\alpha>0}$ as its resolvent. To see that $(P_t^\mu)_{t>0}$ is $m$-symmetric and strongly continuous, we need only show that $(U_\alpha^\mu)_{\alpha>0}$ is so. For $f, g \in L^2(X, m)$, by Lemma 3.1 we have

$$(U_\alpha^\mu f, g) = \mathcal{E}_\alpha^\mu(U_\alpha^\mu f, U_\alpha^\mu g) = (f, U_g^\mu).$$

Hence $U_\alpha^\mu$ is $m$-symmetric. For $f \in C_0(X)$, we have obviously

$$\lim_{\alpha \to \infty} \alpha U_\alpha^\mu f(x) = f(x) \quad \text{m.a.e.}$$

and consequently $(U_\alpha^\mu)_{\alpha>0}$ is strongly continuous on $L^2(X, m)$ (c.f. [F2] Lemma 1.4.3). For proving $(\mathcal{E}^\mu, \mathcal{F}^\mu)$ is the associated Dirichlet form of $(P_t^\mu)$, we first assume that $\mu$ is finite. Let $\mathcal{E}^*, \mathcal{F}^*)$ be the Dirichlet form associated with $(P_t^\mu)_{t>0}$, $\mathcal{F}^\mu$ and $\mathcal{E}^\mu(f, f) = \mathcal{E}^*(f, f)$ Let $f \in \mathcal{F}^\mu$ be a bounded quasicontinuous function, then

$$\alpha(f - \alpha U_\alpha^\mu f, f) = \alpha(f - \alpha U_\alpha f, f) + (E. [\int_0^\infty e^{-\alpha t} \alpha (U_\alpha^\mu f)(X_t) dA_t^\mu, \alpha f) \tag{3.14}$$

From [F2] Lemma 5.1.4 we can prove that the above last term is expressed by

$$< E. [\int_0^\infty e^{-\alpha t} \alpha f(X_t) dt], U_\alpha^\mu(\alpha f) >_\mu . \tag{3.15}$$

Consequently

$$\lim_{\alpha \to \infty} \alpha(f - \alpha U_\alpha^\mu f, f) = \mathcal{E}(f, f) + < f, f >_\mu < \infty, \tag{3.16}$$

which implies $f \in \mathcal{F}^*$ and $\mathcal{E}^*(f,f) = \mathcal{E}^\mu(f,f)$. Conversely if $f \in \mathcal{F}^*$ is nonnegative, then (3.4) implies that $f \in \mathcal{F}$ and hence any element of $\mathcal{F}^*$ admits a quasi continuous modification. Again by (3.14) – (3.16) we see that all the bounded elements of $\mathcal{F}^*$ are elements of $\mathcal{F}^\mu$. Thus by approximating any element in $\mathcal{F}^\mu \cup \mathcal{F}^*$ with bounded elements in $\mathcal{F}^\mu \cap \mathcal{F}^*$, we conclude $(\mathcal{E}^*, \mathcal{F}^*) = (\mathcal{E}^\mu, \mathcal{F}^\mu)$.

We now assume that $\mu$ is a general smooth measure. We can always take an increasing sequence of smooth measures $\{\mu_n\}_{n\geq 1}$ such that $\mu = \sup_n \mu_n$ and $\mu_n(X) < \infty$ for each $n$. It is easy to check that

$$\mathcal{F}^\mu = \{f \in \cap_{n\geq 1}\mathcal{F}^{\mu_n} : \sup_n \mathcal{E}^{\mu_n}(f,f) < \infty\}$$

and

$$\mathcal{E}^\mu(f,f) = \sup_n \mathcal{E}^{\mu_n}(f,f).$$

Moreover, the strong continuity of $(P_t^\mu)$ and Lemma 3.1 imply that $\mathcal{F}^\mu$ is dense in $L^2(X,m)$. Hence by the monotone convergence theorem of forms ([RS] Th S.14) $(\mathcal{E}^\mu, \mathcal{F}^\mu)$ is the closed form associated with $(P_t^\mu)_{t>0}$, which completes the proof. $\square$

**3.3 Remark** (i) Let $\mu$ be a nowhere Radon smooth measure, then $(\mathcal{E}^\mu, \mathcal{F}^\mu)$ is a Dirichlet form such that $\mathcal{F}^\mu$ contains no non-zero continuous functions.
(ii) In the case of $\mu$ being a Radon smooth measure, Oshima proved in [0] Th 4.3.7 that $(\mathcal{E}^\mu, \mathcal{F}^\mu)$ is a regular Dirichlet space associated with $(P_t^\mu)_{t>0}$. His result is also available for non symmetric Dirichlet spaces.
(iii) In the case of $\mu = \mu^+ - \mu^-$ with $\mu^+, \mu^- \in S$, it was proved in [AM3] Th 4.1 that the following assertions are equivalent to each other.
(a) $(P_t^\mu)_{t>0}$ is a strongly continuous semigroup on $L^2(X,m)$.
(b) There exists $\alpha > 0$ such that $U_\alpha^\mu f \in L^2(X,m)$ for all $f \in L^2(X,m)$.
(c) $(\mathcal{E}^\mu, \mathcal{F}^\mu)$ is lower semibounded.
(d) $Q_{\mu^-}$ is relatively form bounded with respect to $(\mathcal{E}^{\mu^+}, \mathcal{F}^{\mu^+})$ with bound $\leq 1$.
Furthermore, if any of the above assertions (a) – (d) holds, then the closed quadratic form corresponding to $(P_t^\mu)_{t>0}$ is the largest closed quadratic form that is smaller than $(\mathcal{E}^\mu, \mathcal{F}^\mu)$.

## 4. General conditions for $(\mathcal{E}, \mathcal{F})$ to be associated with a reasonable Markov process

In this section we assume that $X$ is a metrizable topological space and $m$ is a $\sigma$-finite Borel measure on $X$. Let $(\mathcal{E}, \mathcal{F})$ be a Dirichlet form on $L^2(X, m)$. For a closed set $F \subset X$, we set

$$\mathcal{F}_F = \{f \in \mathcal{F} : f = 0 \text{ m.a.e on } X - F\}. \tag{4.1}$$

**4.1 Definition**　　An increasing sequence of closed sets $\{F_k\}_{k \geq 1}$ is called an $\underline{\mathcal{E}\text{-nest}}$ if $\cup_{k \geq 1} \mathcal{F}_{F_k}$ is $\mathcal{E}_1$-dense in $\mathcal{F}$. A subset $B \subset X$ is said $\mathcal{E}$-polar if there exists an $\mathcal{E}$-nest $\{F_k\}$ such that $B \subset \cap_{k \geq 1}(X - F_k)$. A property is said to hold $\mathcal{E}$-q.e. if it holds except on an $\mathcal{E}$-polar set. Two functions defined $\mathcal{E}$-q.e. on $X$ are said to be $\underline{\mathcal{E}\text{-quasi-equivalent}}$ if they are equal $\mathcal{E}$-q.e.. A function $u$ defined $\mathcal{E}$-q.e. on $X$ is said to be $\underline{\mathcal{E}\text{-quasicontinuous}}$ if there exists an $\mathcal{E}$-nest $\{F_k\}$ such that $u_{|F_k}$ is continuous for each $k$.

An $h$-weighted capacity $\text{Cap}_h$, which is always no bigger than the usual capacity Cap described in Section 2, has been introduced in [AM5]. It has been proved in [AM5] that a set $A$ is an $\mathcal{E}$-polar set if and only if $\text{Cap}_h(A) = 0$. Consequently every nest is an $\mathcal{E}$-nest, and every quasicontinuous function is an $\mathcal{E}$-quasicontinuous function. See also Section 3 for a characterization of $\mathcal{E}$-nests in terms of the capacity Cap. It is also known that each $\mathcal{E}$-polar set is an $m$-negligible set. See [AM5] and the announcement [AM7] for details.

In dealing with Markov processes on a general topological space $X$, we always assume that a cemetery point $\Delta \not\ni X$ is adjoint to $X$ as an isolated point of $X_\Delta := X \cup \{\Delta\}$. Any function $f$ on $X$ is extended to $X_\Delta$ by setting $f(\Delta) = 0$.

**4.2 Definition**　　Let $(X_t) := (\Omega, M, M_t, X_t, \Theta_t, P_x)$ be a strong Markov process with state space $(X, \mathcal{B}(X))$ and life time $\zeta$. $(X_t)$ is called a $\underline{\text{perfect process}}$ if it satisfies the following properties.
(i) Normal property: $P_x(X_0 = X) = 1, \quad \forall x \in X_\Delta$. $\tag{4.2}$
(ii) Cadlag property: $P_x\{w : X_t(w) \text{ is right continuous and has left limit } X_{t-}(w) \text{ on}$
$[0, \zeta(w)] = 1, \forall x \in X_\Delta$ $\tag{4.3}$
(We always put $Z_{0-} = Z_0$ for an arbitrary process $(Z_t)_{t \geq 0}$).
(iii) Regularity of the resolvent: $U_1 f(X_{t-}) I_{\{t < \zeta\}}$ is $P_x$-indistinguishable from $U_1 f(X_t)_-$ $I_{\{t < \zeta\}}, \forall x \in X, f \in \mathcal{B}_b(X)$. ($\mathcal{B}_b(X)$ denotes all the bounded Borel functions on $X$, $U_1 f(x) := E_x[\int_0^\infty e^{-t} f(X_t) dt]$.)

**4.3 Remark**　　Every special standard process, in particular, every Hunt process is a perfect process. See [AM5] for details on this remark.

For an arbitrary subset $A$ of $X_\Delta$, we put

$$\sigma_A = \inf\{t > 0 : X_t \in A\} \tag{4.5}$$

**4.4 Lemma**     Let $(X_t)$ be a perfect process with state space $X$ and life time $\zeta$. Suppose that $X$ is a polish space. Then, there exists an increasing sequence of compact sets $\{F_n\}_{n \geq 1}$ such that

$$P_x\{\lim_{n \to \infty} \sigma_{X-F_n} \geq \zeta\} = 1, \quad \text{m.a.e. } x \in X \tag{4.6}$$

The above lemma can be proved following an idea of [LR], for details see [AMR].

Lemma 4.4 suggests us the following definitions.

**4.4 Definition**     (i) A strong Markov process $(X_t)$ with state space $X$ and life time $\zeta$ is said to be m–tight, if ther exists an increasing sequence of compact sets $\{F_n\}_{n \geq 1}$ such that (4.6) holds.

(ii) A perfect process $(X_t)$ is called an m–perfect process if it is m–tight.

Lemma 4.4 shows that every perfect process $(X_t)$ on $X$ is m–perfect provided $X$ is a polish space.

Recall that a Markov process $(X_t)$ is said to be associated with the Dirichlet form $(\mathcal{E}, \mathcal{F})$, if and only if

$$P_t f = T_t f \quad \text{m.a.e.,} \quad \forall f \in L^2(X, m),$$

where $P_t f(x) = E_x[f(X_t)]$ and $(T_t)_{t>0}$ is the semigroup corresponding to $(\mathcal{E}, \mathcal{F})$. $(X_t)$ is said perfectly associated with $(\mathcal{E}, \mathcal{F})$, if $P_t f$ is an $\mathcal{E}$–quasicontinuous version of $T_t f$ for each $t > 0$ and $f \in L^2(X, m)$.

**4.6 Theorem**     Let $(\mathcal{E}, \mathcal{F})$ be a Dirichlet form on $L^2(X, m)$. Then the following conditions (i) – (iii) are necessary and sufficient conditions for the existence of an m–perfect process $(X_t)$ associated with $(\mathcal{E}, \mathcal{F})$.

(i) There exists an $\mathcal{E}$–nest $\{F_k\}_{k \geq 1}$ consisting of compact sets                (4.7)

(ii) There exists an $\mathcal{E}_1$–dense subset $\mathcal{F}_0$ of $\mathcal{F}$ consisting of $\mathcal{E}$–quasicontinuous

functions                                                                                      (4.8)

(iii) There exists a countable subset $B_0$ of $\mathcal{F}_0$ and an $\mathcal{E}$–polar set $N$ such that $\sigma\{u : u \in B_0\} \supset \mathcal{B}(X) \cap (X - N)$                (4.9)

Moreover, if an m–perfect process $(X_t)$ is associated with $(\mathcal{E}, \mathcal{F})$, then it is always perfectly associated with $(\mathcal{E}, \mathcal{F})$.

The proof of Theorem 4.6 is rather longer. In proving the sufficiency of the conditions we follow the same line of construction of processes used in [F2] chap 6 and employ the quasicontinuous kernels obtained in [AM4]. In proving the necessity of the condition (4.8) we make use of an idea of [L]. See [AM5] for details of the proof.

For a Borel function $f$ on $X$, we put $supp[f] = supp[|f| \cdot m]$. Following [F2], we say that $(\mathcal{E}, \mathcal{F})$ possesses the local property if

$$\text{u,v} \in \mathcal{F}, \; supp[u] \cap supp[v] = 0 \quad \text{implies} \quad \mathcal{E}(u, v) = 0 \tag{4.10}$$

We call an m–perfect process $(X_t)$ with life time $\zeta$ a diffusion if

$$P_x\{X_t \text{ is continuous in } t \in [0, \zeta)\} = 1, \quad \forall x \in X \tag{4.11}$$

Based on Theorem 4.6 and following the argument of [F2] Th. 4.5.1, we can prove the following result.

**4.7 Theorem**      Let $(\mathcal{E}, \mathcal{F})$ be a Dirichlet form on $L^2(X, m)$. Then the conditions (4.7)–(4.9) together with the local property (4.10) are the necessary and sufficient conditions for the existence of a diffusion process associated with $(\mathcal{E}, \mathcal{F})$.
For the detailed proof of Th. 4.7 see [AM6].                    ∎

## 5. Process associated with $(\mathcal{E}^\mu, \mathcal{F}^\mu)$

Consider again a regular Dirichlet form $(\mathcal{E}, \mathcal{F})$ on a locally compact metrizable space $X$ with a Radon measure $m$. We shall freely use the notations of the sections 2–4. Let $\mu$ be a smooth measure and $(\mathcal{E}^\mu, \mathcal{F}^\mu)$ be defined by (3.4) and (3.5). In Theorem 3.2 we have proved that $(\mathcal{E}^\mu, \mathcal{F}^\mu)$ is a Dirichlet form. In this section we shall show that $(\mathcal{E}^\mu, \mathcal{F}^\mu)$ satisfies the conditions (4.7)–(4.9). Therefore there always exists an $m$–perfect process associated with $(\mathcal{E}^\mu, \mathcal{F}^\mu)$.

For a closed set $F$ and $\alpha > 0$, we set

$$U_\alpha^{\mu, F} f = E_x \Big[ \int_0^{\sigma(X-F)} e^{-\alpha t - A_t^\mu} f(X_t) dt \Big] \tag{5.1}$$

provided the right hand side makes sense.

**5.1 Lemma**      Let $F$ be a closed set and $\alpha > 0$. Then for each $f \in L^2(X, m)$, we have $U_\alpha^{\mu, F} f \in \mathcal{F}_F^\mu$ and

$$\mathcal{E}_\alpha^\mu(U_\alpha^{\mu, F} f, u) = (f, u) \quad , \quad \forall u \in \mathcal{F}_F^\mu \tag{5.2}$$

Morover, $U_\alpha^{\mu, F} f$ is quasicontinuous. Here $\mathcal{F}_F^\mu$ is defined by (4.1) with respect to $\mathcal{F}^\mu$.

**Proof**      Let us put for $n \geq 1$,

$$\mu_n = \mu + n(I_{X-F} \cdot m), \tag{5.3}$$

Then the corresponding PCAF is

$$A_t^{\mu_n} = A_t + n \int_0^t I_{X-F}(X_s) ds \tag{5.4}$$

Without loss of generality we may assume that $f$ is nonnegative. Then $U_\alpha^{\mu_n} f \downarrow U_\alpha^{\mu, F} f$ q.e. (recall $U_\alpha^{\mu_n} f$ is defined by (3.3)). Applying Fubini's theorem we get

$$U_\alpha^{\mu_n} f = U_\alpha^\mu f - U_\alpha^\mu [n(U_\alpha^{\mu_n} f) I_{X-F}] \tag{5.5}$$

Thus by Lemma 3.1 we have $U_\alpha^{\mu_n} f \in \mathcal{F}^\mu$ and

$$\mathcal{E}_\alpha^\mu(U_\alpha^{\mu_n} f, U_\alpha^{\mu_k} f) = (f, U_\alpha^{\mu_k} f) - (n(U_\alpha^{\mu_n} f)I_{X-F}, U_\alpha^{\mu_k} f) \qquad (5.6)$$

Applying Lemma 3.1 again we obtain

$$(n(U_\alpha^{\mu_n} f)I_{X-F}, U_\alpha^{\mu_k} f) = \mathcal{E}_\alpha^{\mu_k} (U_\alpha^{\mu_k} [n(U_\alpha^{\mu_n} f)I_{X-F}], U_\alpha^{\mu_k} f)$$
$$= (U_\alpha^{\mu_k} [n(U_\alpha^{\mu_n} f)I_{X-F}], f) \qquad (5.7)$$

Writing $B_t = \int_0^t I_{X-F}(X_s)ds$, by the Markovian property and Fubini's theorem we have,

$$U_\alpha^{\mu_k} [n(U_\alpha^{\mu_n} f)I_{X-F}] (x)$$
$$= E_x \left[ \int_0^\infty e^{-\alpha t - A_t^\mu} f(X_t)(e^{-nB_t} \int_0^t e^{-\alpha s - A_s^\mu - (k-n)B_s} ndB_s)dt \right] \qquad (5.8)$$

Since

$$1 \geq e^{-nB_t} \int_0^t e^{-\alpha s - A_s^\mu - (k-n)B_s} ndB_s \to 0 \quad \text{when } k \geq n \to \infty,$$

we conclude from (5.7) and (5.8) that

$$(n(U_\alpha^{\mu_n} f)I_{X-F}, U_\alpha^{\mu_k} f) \to 0, \quad \text{when } k \geq n \to \infty.$$

Therefore from (5.6) we see that $\{U_\alpha^{\mu_n} f\}_{n \geq 1}$ forms an $\mathcal{E}_\alpha^\mu$-Cauchy sequence which implies $U_\alpha^{\mu,F} f \in \mathcal{F}^\mu$, and $U_\alpha^{\mu,F} f$ is quasicontinuous. Moreover, from (5.5) we see that

$$\mathcal{E}_\alpha^\mu(U_\alpha^{\mu_n} f, u) = (f, u) , \quad \forall u \in \mathcal{F}_F^\mu.$$

Letting $n \to \infty$ we obtain (5.2). The proof is completed. $\qquad \square$

**5.2 Remark** Let us put

$$U_\alpha^F f(x) = E_x \left[ \int_0^{\sigma_{X-F}} e^{-\alpha t} f(X_t)dt \right]. \qquad (5.9)$$

By letting $\mu = 0$ in Lemma 5.1 we see that

$$U_\alpha^F f \in \mathcal{F}_F, \quad \forall f \in L^2(X, m), \qquad (5.10)$$

and

$$\mathcal{E}_\alpha(U_\alpha^F f, u) = (f, u), \quad \forall u \in \mathcal{F}_F, f \in L^2(X, m). \qquad (5.11)$$

The above facts were first discovered in the proof of [FO] Lemma 2.1.

**5.3 Proposition**    Let $\{F_n\}_{n\geq1}$ be an increasing sequence of closed sets. Then the following assertions are equivalent to each other.

(i) $\lim_{n\to\infty} \text{Cap}(K - F_n) = 0$ ∀ compact sets $K$ (5.12)

(ii) $P_x\{\lim_{n\to\infty} \sigma_{X-F_n} \geq\} = 1$,    q.e. $x \in X$ (5.13)

(iii) $\{F_n\}_{n\geq1}$ is an $\mathcal{E}$ − nest. (5.14)

(iv) $\{F_n\}_{n\geq1}$ is an $\mathcal{E}^\mu$ − nest. (5.15)

**Proof**    For the proof of (i) ⟺ (ii) see [Fu] Lemma 5.1.6. The equivalence between (ii) and (iii) is obtained from (ii) ⟺ (iv) by setting $\mu \equiv 0$. Thus we need only to prove (ii) ⟺ (iv). Let $\{F_n\}_{n\geq}$ be an increasing sequence of closed sets. We put

$$\sigma := \lim_{n\to\infty} \sigma_{X-F_n}\,,$$

$$\mathcal{U}_n f := U_1^{\mu,F_n} f = E_x[\int_0^{\sigma_{X-F_n}} e^{-t} f(X_t) dt]\,,$$

$$\mathcal{U}_\infty f := E_x[\int_0^\sigma e^{-t} f(X_t) dt]\,,$$

$$\mathcal{U}f := E_x[\int_0^\infty e^{-t} f(X_t) dt]\,,$$

$$H_n f := \mathcal{U}f - \mathcal{U}_n f\,,\quad 1 \leq n \leq \infty\,,$$

provided the right hand sides make sense. Let $f \in L^2(X, m)$. Then the following facts can be derived from Lemma 5.1:

$$\mathcal{U}_n f \in \mathcal{F}_{F_n}^\mu\,; \tag{5.16}$$

$$\mathcal{E}_1^\mu(H_n f, g) = 0\,,\ \forall g \in \mathcal{F}_{F_n}^\mu\,; \tag{5.17}$$

$$\mathcal{U}_n f \to \mathcal{U}_\infty f,\ \text{q.e. } x \in X \text{ and in } \mathcal{E}_1^\mu - \text{norm}; \tag{5.18}$$

$$H_n f \to H_\infty f,\ \text{q.e. } x \in X \text{ and in } \mathcal{E}_1^\mu - \text{norm}. \tag{5.19}$$

Suppose that (ii) holds. Then $\mathcal{U}_\infty f = \mathcal{U}f$ q.e. and (iv) follows from (5.16) and (5.18). Conversely suppose that (iv) is true. Then from (5.17) and (5.19) we see that $H_\infty f = 0$ m.a.e. and hence $H_\infty f = 0$ q.e. because one can show that $H_\infty f$ is quasicontinuous. In particular, if we choose $f \in L^2(X, m)$ to be strictly positive everywhere, then $H_\infty f = 0$ q.e. implies (ii). The proof is completed.    □

We can now prove the following results.

**5.4 Theorem**    (i) There always exists an $m$−perfect process associated with $(\mathcal{E}^\mu, \mathcal{F}^\mu)$. (ii) Suppose that $(\mathcal{E}, \mathcal{F})$ possesses the local property (4.10), then there always exists a diffusion process associated with $(\mathcal{E}^\mu, \mathcal{F}^\mu)$.

**Proof**     (i) We should verify that $(\mathcal{E}^\mu, \mathcal{F}^\mu)$ satisfies the conditions (4.7)–(4-9). Let $\{F_n\}_{n\geq 1}$ be an increasing sequence of compact sets satisfying conditions (2.4)–(2.6). Then by Lemma 5.3 $\{F_n\}_{n\geq 1}$ is an $\mathcal{E}^\mu$–nest. Hence $(\mathcal{E}^\mu, \mathcal{F}^\mu)$ satisfies (4.7). Also by Lemma 5.3 every quasicontinuous function is $\mathcal{E}^\mu$–quasicontinuous. Thus $(\mathcal{E}^\mu, \mathcal{F}^\mu)$ satisfies (4.8) because every element of $\mathcal{F}^\mu$ admits a quasicontinuous modification. To verify the condition (4.9), we take a countable set $\{f_n\}_{n\geq 1}$ of continuous functions with compact supports, such that $\mathcal{B}(X) = \sigma\{f_i : i \geq 1\}$. Let $N$ be an exceptional set of $(A_t^\mu)$ (c.f. Section 3). We put

$$f_{i,j}(x) = \begin{cases} E_x[j \int_0^\infty e^{-jt - A_t^\mu} f_i(X_t)dt] & , x \in X - N \\ f_i(x) & , x \in N \end{cases}$$

Then $f_{ij}(x) \to f_i(x)$ pointwise as $j \to \infty$. By Lemma 3.1 and Lemma 5.3 we see that $B_0 := \{f_{ij} : i \geq 1, j \geq 1\}$ is an $\mathcal{E}^\mu$–quasicontinuous subset of $\mathcal{F}^\mu$. Obviously we have $\sigma\{f_{ij} : f_{ij} \in B_0\} \supset \mathcal{B}(X)$.
Thus (4.9) is verified. The proof is completed by applying Theorem 4.6.

   (ii) We need only to show that $(\mathcal{E}^\mu, \mathcal{F}^\mu)$ possesses the local property (4.10) it $(\mathcal{E}, \mathcal{F})$ does so. But this is evident from the definition (3.5).                                        □


# 6. Application to the quantum theory

Let $m$ be the Lebesgue measure on $\mathbb{R}^d, d \geq 2$. Consider the classical Dirichlet form $(\mathcal{E}, \mathcal{F})$ associated with the Laplacian $-\frac{1}{2}\Delta$ on $\mathbb{R}^d$. More precisely, $\mathcal{F} := H^1(\mathbb{R}^d)$ is the 1-order Sobolev space and

$$\mathcal{E}(f, g) = \frac{1}{2}\int \nabla f \cdot \nabla g m(dx) \quad , \quad \forall f, g \in \mathcal{F} \tag{6.1}$$

As we have mentioned after the proof of Theorem 2.1, there always exists a nowhere Radon smooth measure in this case. Let us fix a nowhere Radon smooth measure $\mu$ and consider the perturbed Dirichlet form $(\mathcal{E}^\mu, \mathcal{F}^\mu)$ defined by (3.4) and (3.5). Since $(\mathcal{E}^\mu, \mathcal{F}^\mu)$ is again a Dirichlet form on $L^2(\mathbb{R}^d, m)$, there is a nonnegative self-adjoint operator, which we denote by $H^\mu$ with the domain $\mathcal{D}(H^\mu)$, such that $(\mathcal{E}^\mu, \mathcal{F}^\mu)$ is the unique quadratic form associated with $H^\mu$.


### 6.1 Proposition
(i) $\mathcal{D}(H^\mu) \subset H^1(\mathbb{R}^d) \cap L^2(\mathbb{R}^d, \mu) \cap L^1_{\text{loc}}(\mathbb{R}^d, \mu)$          (6.2)
(ii) $H^\mu u = (-\frac{\Delta}{2} + \mu)u$    in the following distributional sense:
    $(H^\mu u, \varphi) = (u, -\frac{\Delta}{2}\varphi) + <u, \varphi>_\mu$ ,   $\forall u \in \mathcal{D}(H^\mu), \varphi \in \mathbb{C}_0^\infty(\mathbb{R}^d)$     (6.3)

**Proof**    Let $U_\alpha^\mu f$ be defined by (3.3). It is well known that

$$\mathcal{D}(H^\mu) = \{U_\alpha^\mu f: \quad f \in L^2(\mathbb{R}^d, m), \quad \alpha > 0\} \tag{6.4}$$

and for $U_\alpha^\mu f \in \mathcal{D}(H^\mu)$,

$$H^\mu(U_\alpha^\mu f) = f - \alpha U_\alpha^\mu f \tag{6.5}$$

By Lemma 3.1 we see that $\mathcal{D}(H^\mu) \subset H^1(\mathbb{R}^d) \cap L^2(\mathbb{R}^d, \mu)$.
Noticing that every function $\varphi \in C_0^\infty(\mathbb{R}^d)$ can be expressed by the difference of two $\gamma$-excessive functions, we see from (3.11) that

$$\mathcal{E}_\alpha(U_\alpha^\mu f, \varphi) = (f, \varphi) - <U_\alpha^\mu f, \varphi>_\mu, \quad \forall \varphi \in C_0^\infty(\mathbb{R}^d) \tag{6.6}$$

From (6.6) we see that $\mathcal{D}(H^\mu) \subset L_{\mathrm{loc}}^1(\mathbb{R}^d, \mu)$. From (6.6) and (6.5) we conclude (6.3), which completes the proof.                                                    $\square$

The above proposition shows that $H^\mu$ is a Schrödinger operator with potential $\mu$. The Schrödinger operator $H^\mu$ is singular on each neighboughood of every point $x \in \mathbb{R}^d$, since $\mu$ is nowhere Radon. Some visual examples of such singular examples can be constructed as follows.

**6.2 Examples**    Let $\{x_j\}_{j\geq 1}$ be a dense subset of $\mathbb{R}^d$ and let $\{\alpha_j\}_{j\geq 1}$ be a sequence of real numbers with $\alpha_j \geq d, \forall j \geq 1$. Then there always exists a sequence of strictly positive numbers $\{c_j\}_{j\geq 1}$ such that $f \cdot m$ is a nowhere Radon smooth measure, where $f(x)$ is given by

$$f(x) := \sum_{j\geq 1} c_j |x - x_j|^{-\alpha_j} \tag{6.7}$$

The corresponding Schrödinger operator $H^\mu := \frac{-\Delta}{2} + f$ has a potential singular on each non-empty open set.

For details of the above example see [AM2] Proposition 1.3.

Applying Theorem 5.4, we see that there always exists a diffusion process associated with the Schrödinger operator $H^\mu$, even though the Schrödinger operator is singular on each neighboughood of every point $x \in \mathbb{R}^d$ and the corresponding form domain contains no continuous functions (except for the zero-function).

## Acknowledgements

It is a great pleasure to thank Prof. Dr. Ana Bela Cruzeiro for her very kind invitation to attend a most stimulating conference. We are very gratefule to Professors Masatoshi Fukushima and Michael Röckner for stimulating comments and discussions. Partial financial support by the SFB 237 (Essen-Bochum-Düsseldorf) is also gratefully acknowleged.

# References

[AH1] S. Albeverio, R. Høegh-Krohn: *Quasi-invariant meausures, symmetric diffusion processes and quantum fields.* In: Les méthodes matématiques de la théorie quantitique des champs, Colloques Internationaux du C.N.R.S., no. **248**, Marseille, 23–27 juin 1975, C.N.R.S., 1976.

[AH2] S. Albeverio, R. Høegh-Krohn: *Dirichlet forms and diffusion processes on rigged Hilbert spaces.* Z. Wahrscheinlichkeitstheorie verw. Gebiete **40**, 1–57 (1977).

[AH3] S. Albeverio, R. Høegh-Krohn: *Hunt processes and analytic potential theory on rigged Hilbert spaces.* Ann. Inst. Henri Poincar'e, vol. XIII, no. 3, 269–291 (1977).

[AM1] S. Albeverio, Zhiming Ma: *Nowhere Radon smooth measures, perturbations of Dirichlet forms and singular quadratic forms*, pp. 3-45 in Proc. Bad Honnef Conf. 1988, Edts. N. Christopeit, K. Helmes, M. Kohlmann, Lect. Notes Inf. Control (126), Springer (1989)

[AM2] S. Albeverio, Zhiming Ma: *Additive functionals, nowhere Radon and Kato class smooth measures associated with Dirichlet forms*, BiBoS Nr. 395 (1988)

[AM3] S. Albeverio, Zhiming Ma: *Perturbation of Dirichlet forms – lower semiboundedness, closability and form cores*, SFB 237 preprint Nr. 67 (1989), to appear in J. Functional Analysis

[AM4] S. Albeverio, Zhiming Ma: *A note on quasicontinuous kernels representing quasi-linear positive maps.* BiBoS preprint (1990)

[AM5] S. Albeverio, Zhiming Ma: *Necessary and sufficient conditions for the existence of m-perfect processes associated with Dirichlet forms*, BiBoS preprint (1990)

[AM6] S. Albeverio, Zhiming Ma: *Local property for Dirichlet forms on general metrizable spaces*, in preparation

[AM7] S: Albeverio, Zhiming Ma: *A general correspondence between Dirichlet forms and right processes*, SFB 237-Preprint (1990)

[AMR] S. Albeverio, Zhiming Ma, M. Röckner: Paper in preparation

[AR1] S. Albeverio, M. Röckner: *Classical Dirichlet forms on topological vector spaces – construction of an associated diffusion process.* BiBoS preprint Nr. 342 (1988), to appear in Prob. Th. Rel. Fields.

[AR2] S. Albeverio, M. Röckner: *Stochastic differential equations in infinite dimensions: solutions via Dirichlet forms*, SFB 237-Preprint (1990)

[AR3] S. Albeverio, M. Röckner: *Infinite dimensional diffusions connected with positive generalized white noise functionals*, to appear in Proc. Bielefeld Conf. "White Noise Analysis", Ed. T. Hida, H.H. Kuo, J. Potthoff, L. Streit

[BM1] Ph. Blanchard, Zhiming Ma: *Semigroup of Schrödinger operators with potentials given by Radon mmeasures*, to appear in "Stochastic Processes – Physics and Geometry, Edts. S. Albeverio, G. Casati, U. Cattaneo, D. Merlini, R. Moresi, World Scient., Singapore (1989)

[BM2] Ph. Blanchard, Zhiming Ma: *Smooth measures and Schrödinger semigroups*, BiBoS 295 (1987)

[BM3] Ph. Blanchard, Zhiming Ma: *New results on the Schrödinger semigroups with potentials given by signed smooth measures*, to appear in Proc. Silivri Workshop 1988, Ed. Korzlioglu et al., Lect. Notes Maths., Springer (1988)

[F1] M. Fukushima: *Dirichlet spaces and strong Markov processes*. Trans. Amer. Math. Soc. **162**, 185–224 (1971).

[F2] M. Fukushima: *Dirichlet forms and Markov processes*. Amsterdam-Oxford-New York: North Holland 1980.

[F3] M. Fukushima: *Basic properties of Brownian motion and a capacity on the Wiener space*. J. Math. Soc. Japan **36**, 161–175 (1984).

[FO] M. Fukushima, Y. Oshima: *On skew product of symmetric diffusion processes*, Forum Math. 1 (1989), 103–142.

[K] S. Kusuoka: *Dirichlet forms and diffusion processes on Banach space*, J. Fac. Science Univ. Tokyo, Sec. 1A **29**, 79–95 (1982)

[L] Y. Le Jan: *Quasi-continuous functions and Hunt processes*, J. Math. Soc. Japan, **25**, 37–42 (1983)

[LR] T. Lyons, M. Röckner: *A note on tightness of capacities associated with Dirichlet forms*, Preprint (1989)

[O] Y. Oshima: Lecture on Dirichlet forms, Erlangen (1988)

[RS] M. Reed, B. Simon: *Methods of Modern Mathematical Physics (II)*, reviesed and enlarged edition, Academic Press New York (1980)

[S] M.L. Silverstein: *Symmetric Markov Processes*, Lect. Notes Maths. **426**, Springer Berlin (1974)

# Geometric Quantization on Wiener Manifolds

Lars Andersson, Gunnar Peters

Department of Mathematics

Royal Institute of Technology

S-10044 Stockholm

Sweden

**Abstract.** The Geometric Quantization procedure is considered in the context of Wiener manifolds.

**Introduction.** The material presented here may be viewed as a continuation of the discussion in [1] where some basic aspects of the problem of extending the geometric quantization procedure to the infinite dimensional case were discussed.

In [3][24][25] the problem of quantizing a linear phase space has been considered by taking the limit of results obtained in finite dimensional geometric quantization. Many of these objects, e.g. the trace of certain linear mappings, the Laplace operator and Kähler potentials, do not have a well defined limit as $n$ goes to infinity and have to be taken care of by hand. Here we construct a geometric quantization-procedure on the infinite dimensional phase space and obtain Bargmann spaces, the Bogoliubov transformation and infinitesimal pairing expressed in terms of well defined objects on a Wiener space.

Let $H$ be a (strongly) symplectic Hilbert space with a complex structure. Given a Lagrangian subspace $W_0$ in $H^{\mathbf{C}}$, the restricted Siegel upper half plane $K_+$ is defined as the set of $AW_0$ for $A \in Sp_r$, the restricted symplectic group. The vacuum in the Fock space $S'(W)$ defined w.r.t. $W$ defines a line bundle $\delta$ over $K_+$. This line bundle plays the role in our work of the bundle of half forms which occurs in the BKS pairing in the finite dimensional case. As discussed in [16] a $U(1)$ extension $Mp_r$ of $Sp_r$ acts on $\delta$.

The arena for GQ in the infinite dimensional case considered here is Wiener manifolds. We use a technical setup based on the work of Kusuoka [10]. This may be expected to generalize further to a setup based entirely on the concept of smoothness defined in terms of the Frechet spaces $W_\infty$ used in the Malliavin calculus.

In contrast to eg. the definition of Wiener manifold used in the work of Piech [13] and Eells-Elworthy [7] the definition given below does *not* imply that a Wiener manifold $M$ is a Banach manifold. Rather we consider regularity conditions defined purely in terms of measurability and regularity in $H$-directions. In this setting, a Wiener manifold has a natural tangent bundle with Hilbert space structure on the fibers (but no tangent bundle with fibers modelled on $B$) and a $Gl_r$ reduction of the frame bundle.

Supported in part by NFR, the Swedish Academy of Sciences and the Gustavsson foundation

Typeset by $\mathcal{A}_{\mathcal{M}}\mathcal{S}$-TEX

It should be stressed that the definition of Wiener manifold used in this paper is in a rather preliminary form and a number of technical issues need further study. In particular, the existence of such structures as a partition of unity and a Poincare Lemma in this context need careful consideration. We do not expect this to present any serious problems. Since all of the material presented here can be made rigorous using more restrictive assumptions like in [13], we concentrate on the formalism in the present note and leave a detailed study of the technical issues raised to a later paper.

A symplectic structure reduces the $Gl_2$ structure on the frame bundle to an $Sp_2$ structure which extends to a $Mp_2$ structure. Given a polarization $W_0$ on $M$ (a Lagrangian subbundle of $T^{\mathbb{C}}M$, a vacuum line bundle $\delta$ is defined using the above mentioned construction on each fiber of $D(M; W_0)$, the bundle over $M$ with fiber given by the restricted Siegel upper half plane w.r.t. $W_{0,x}$.

We introduce a prequantization representation on a space of section of a prequantum bundle $B \otimes W^{1/2}$ over $M$ where $B$ is a Hermitian line bundle and $W^{1/2}$ is a Wiener bundle like in [2]. The BKS representation of the observables is now constructed as in [23], in the sense that given a polarization on a Wiener manifold we construct a representation Hilbert space out of polarized sections of the prequantum bundle, use the prequantum representation to act on these polarized wave functions and then correct the pairing by tensoring with the vacuum bundle $\delta$ pulled back to $M$ by the section $W$ of the Lagrangian frame bundle.

In the infinite dimensional case, there have been a few discussions of issues involving concepts from the GQ construction, see [2], [23] (scalar fields), [26] (quantum gravity), [5] (string theory). None of these contain either a complete geometrical setup or a consideration of the technical issues involved.

We do not expect the present setup to be directly applicable to any interacting field theory, even in the 2-dimensional case, except after a momentum cutoff. It is an interesting problem to apply the present techniques to an integrable field equation such as the Sine-Gordon equation. Presumably, in this case, the problem of analyzing the stochastic extensions of classical objects to the Wiener version of the phase space is tractable.

## 1 Technical Preliminaries.

We give a very sketchy introduction to the technical setting which will be used in the following discussion. As stated above, we do not at present have a complete picture of this, but concentrate on formal aspects of the constructions in this paper.

### 1.1 Analysis on Wiener Space.

**Sobolev Spaces** Let $(B, H, \mu)$ be an AWS. Then the Sobolev spaces $W_{p,k}$ and the Frechet space $W_\infty$ are defined (Kree, Malliavin). Functions in $W_\infty$ are measurable and continuous quasi-a.e. w.r.t. the capacities $C_{p,r}$ but are *not* continuous w.r.t. the $B$-topology.

**Local Sobolev Spaces** Let $U$ be a sufficiently regular measurable set ($H - C^1$ open [10]) in $B$. Then the Sobolev spaces $W_{p,r}(U)$ and the Frechet space $W_\infty(U)$ are defined. Further we need the concept of $H - C^k$ mapping for $k \in \mathbb{N} \cup \infty$ defined as in [10]. Note that the definition of $H - C^k$ mapping used here does not imply continuity w.r.t. the $B$-topology, but only measurability.

**Nonlinear Transformations of Wiener Measure** Let $\phi$ denote a measurable mapping $B \to B$. Then $\phi^* \mu$ is absolutely continuous w.r.t. $\mu$ on a $H - C^1$-open set $U \subset B$ if $\phi - I : U \to H$ is $H - C^1$. For $\phi$ such that $\phi - I \in H - C^\infty$ we denote the Jacobian function by

$$J_\phi = \frac{d\phi^* \mu}{d\mu}.$$

Then $J_\phi \in W_\infty$.

## 1.2 Wiener manifolds.

**DEFINITION (WIENER MANIFOLD).** *Let $(M,\sigma)$ be a measure space with $\sigma$-algebra $\sigma$ and $\{V_\alpha\}$ a countable locally finite set of generators of $\sigma$. If there exists mearurable bijections $\phi_\alpha: V_\alpha \to U_\alpha \subset B$ such that $U_\alpha$ are $H$-$C^k$ and whenever $V_\alpha \cap V_\beta$ is nonempty $\phi_\alpha \phi_\beta^{-1} - I \in H$-$C^\infty( \phi_\beta(V_\alpha \cap V_\beta) \to H)$ and if there exists a $W_\infty$ partition of unity on $M$ subordinate to $\{V_\alpha\}$ then we say that $\{V_\alpha, \phi_\alpha\}$ is a Wiener atlas for $M$ and that $(M,\{V_\alpha, \phi_\alpha\})$ is a Wiener manifold.*

We expect to be able to prove the existence of a $W_\infty$ partition of unity on $M$ from the remaining properties of a Wiener manifold.

We denote $J_{\alpha\beta} = J_{\phi_\alpha \phi_\beta^{-1}}$.

The assumptions on the transition functions allow us to define a tangent space $TM_x$ modelled on H at each point x of M. We will use the notation $TM = \cup \, TM_x$. There is also a naturally defined cotangent bundle $T^*M$. TM has a natural $Gl_2(H)$ reduction. Note that there is no concept of a tangent bundle with fibres isomorphic to B in this setting.

REMARK. The interesting properties of the transition functions $\phi_\alpha \phi_\beta^{-1}$ are that they preserve the $W_\infty$ structures and the equivalence class of the Gaussian measure on B. We propose further investigation of this topic. In this paper we keep the property of point wise defined functionals explicit in the definition of H-$C^\infty$, which enables us to define the tangent bundle TM as above.

REMARK. An immediate example of a nonlinear Wiener manifold is $B \times N$, where N is a finite dimensional Riemannian manifold. If $\{W_\alpha, \psi_\alpha\}$ is a smooth locally finite atlas for N then $\{B \times W_\alpha, I \times \psi_\alpha\}$ is a Wiener atlas for $B \times N$.

**DEFINITION (WIENER BUNDLES).** *By [7] the $J_{\alpha\beta}s$ define a real line bundle W on M. This bundle is trivial and we can define powers $W^r$ of it for all $r \in R$.*

On the $V_\alpha$s we define measures $\varepsilon_\alpha = \varphi_\alpha^* \mu$. Any section w of $W^r$ such that the local representatives $w_\alpha$ are in $L_{1/r}(\varepsilon_\alpha)$ defines a measure $\mu_s$ on M by

$$\mu_s(O) = \int_O w_\alpha^{1/r} d\varepsilon_\alpha \text{ for } O \subset V_\alpha. \qquad (1)$$

Let K be a separable Hilbert space. The differential D of a K valued functional is defined on $V_\alpha$ through pullback via $\varphi_\alpha$ of the H-differential on B. It is then possible to define local Sobolev norms on $V_\alpha$ using the measure $\varepsilon_\alpha$. The corresponding Sobolev spaces will be denoted $W_{p,n}(V_\alpha,K)$. There is a $W_\infty$ structure on M using the local Sobolev norms and the partition of unity. We say that a functional is locally $W_\infty$ if the restriction to any $V_\alpha$ is in $W_\infty(V_\alpha,K)$. If $f \in W_\infty(K)$ then $Df \in W_\infty(TM \otimes K)$.

The local dual $D_\alpha{}^*$ of D w.r.t. $\varepsilon_\alpha$ is defined through the relation

$$\int_{U_\alpha} <g,D_\alpha{}^*f>_K d\varepsilon_\alpha = \int_{U_\alpha} <Dg,f>_{HS} d\varepsilon_\alpha , \qquad (2)$$

where $g \in W_\infty(V_\alpha,K)$ has support in $V_\alpha$ and $f \in W_\infty(V_\alpha,TM \otimes K)$. The inner product $<,>_{HS}$ is defined in [10] and depends on the choice of inner product on the fibers of TM. In the following it will be assumed that a global Riemannian structure has been choosen on TM.

**Lemma 1:** *Fix $\alpha$. For each $g \in W_\infty(V_\alpha,C)$ and $X \in W_\infty(V_\alpha,TM \otimes C)$ we have that*

$$D^*_\alpha(gX) = gD^*_\alpha(X) - Dg(X). \qquad (3)$$

*Proof:* Test both sides against a functional f with support in $V_\alpha$.

**Lemma 2:** $D^*_\alpha(X) = D^*_\beta(X) - Dln(J_{\alpha\beta})(X)$ *on* $V_\alpha \cap V_\beta$. $\qquad (4)$

*Proof:* Test both sides against a functional f with support in $V_\alpha$ and use lemma 1.

Vector fields X will be $W_\infty$ sections of TM. This implies that X is locally a $W_\infty$ function from $V_\alpha$ to H. A Riemannian structure g on M is a choice of inner product on the fibres of TM compatible with those induced by the $\phi_\alpha$s such that g(X,Y) is well defined and in $W_\infty(M)$ for X and Y in $W_\infty(TM)$.

REMARK . Note that it is a nontrivial question wether or not there exist local flows for vector fields in the above sense. On this point see [6].

Following [19] we define local Sobolev norms for sections of tensor bundles $T_s^r$ of TM locally isomorphic to the topological tensor product $H \otimes ... \otimes H^*$. As before it is possible to define a global $W_\infty$ structure for $T_s^r$. Smooth n-forms will be alternating sections in $W_\infty(T_n)$. Smooth n-forms on M will be denoted $\Lambda^n(M)$.

**Poincaré Lemma.** A Poincaré Lemma for globally defined forms on a Wiener space follows from the work in [19]. We need a local version, see [13] for a proof using stronger technical assumptions than here. It is to be expected that this generalizes to the present situation. This problem will be considered in a future paper.

We will also need objects which do not obey the Hilbert Schmidt conditions implicit in the constructions of Shigekawa. Let $L^rM$ be the bundles of bounded multilinear $\mathbf{R}$ valued mappings on the fibres of TM. To define smoothness of a section of $L^rM$ it is necassary to choose some extra background structure on H.

**DEFINITION(WILD STRUCTURES).** *Let $S_0$ be a polynomial mapping $H \rightarrow L^r(H)$. Having choosen an $S_0$ we say that a section S of $L^rM$ is wild if and only if $S - \varphi_\alpha {}^*S_0$ extends to a section over $V_\alpha$ which is locally in $W_\infty(T^r)$.*

For this definition to be consistent $\varphi_\alpha {}^*S_0 - \varphi_\beta {}^*S_0$ has to be in $W_\infty(T^r)$ for all pairs $(\alpha,\beta)$. This is an extra condition of the Wiener atlas and we hope to give more precise information about when this is true in future papers.

An important example is when S takes values in the set of alternating multilinear functionals on the fibres of TM. Such a section will be called a wild n-form. Let $S_0$ be as above but with values in the alternating elements of $L^n(H)$. Using the Frechét derivative on H it is possible to define the exterior derivative of $S_0$. It can easily be proved that $dS_0$ is an n+1 form on H and that $d\circ d = 0$. In this context we have a Poincaré Lemma.

Since a wild n-form S on M is such that $S - \varphi_\alpha {}^*S_0$ is locally in $W_\infty(T^r)$ we define

$$dS = d(S - \varphi_\alpha {}^*S_0) + \varphi_\alpha {}^*dS_0. \qquad (5)$$

If $dS_0 = 0$ and there exists a $\theta_0$ such that $d\theta_0 = S_0$ then a Poincaré Lemma on M for Shigekawa forms implies a Poincaré Lemma for wild forms on M. Examples of wild structures are the Riemannian structure g, the symplectic fundamental form $\omega$, see below, and the symplectic potential if it exists.

## 2 Symplectic Wiener manifold.

Given a strong symplectic form $\omega_0$ and a symplectic potential $\theta_0$ on H we make the following definition.

**DEFININITION(SYMPLECTIC WIENER MANIFOLD).** *A symplectic Wiener manifold $(M,\omega)$ is a Wiener manifold M together with a wild, w.r.t. $\omega_0$, closed 2-form which is a strong symplectic form on each fibre of TM.*

Since $\omega_0$ is a constant 2-form on H it presents no problems. $\theta_0$ and the Wiener atlas is supposed to be such that the conditions mentioned after the definition of wild structures are satisfied. In this case the discussion in the last section shows that a Poincaré Lemma for smooth n-forms on M gives us local symplectic potentials $\theta$ for $\omega$.

Formally an $f \in W_\infty(M)$ generates a hamiltonian vector field $X_f$ through the relation

$$df + C_{X_f}\omega = 0,$$

where C denotes contraction. The question when $X_f$ exists as an element of $W_\infty(TM)$ and if so when $X_f$ generates a local flow that preserve the $W_\infty$ structure on M and the equivalence class of the local measures $\varepsilon_\alpha$ has to be considered. We say that $X_f$ is complete if it generates a complete flow $\gamma_t$.

A quantization of $(M,\omega)$ is a representation of some subalgebra of $W_\infty(M)$ as selfadjoint operators on a Hilbert space.

REMARK. To quantize for example a free field theory it is necassary to extend the representation to some observables not in $W_\infty(M)$. These observables will in general be such that they generate unitary flows on M w.r.t. some Riemannian structure g and it is expected that the extension will exist, see section 7.6 .

## 3 Prequantization.

We now set out to construct a prequantum bundle P over $(M,\omega)$ such that $P \otimes P^* = W^1 \otimes C$. It is then possible to define an $L_2$ structure on P by the inner product $<s,t> = \mu_{st}*(M)$, see equation (1). $\mu_{st}*$ is here a complex measure absolutely continuous w.r.t. the background measures $\varepsilon_\alpha$. The observables will be represented as differential operators on P which are formally symmetric w.r.t. $<,>$.

Let $\{(V_\alpha,\varphi_\alpha)\}$ be a Wiener atlas such that the Poincaré Lemma is valid on all possible intersections of the $V_\alpha$s. It is expected that the Poincaré Lemma is valid on topologicaly trivial sets but stronger assumptions on the regularity of the sets might be needed. As we have seen above there exists local symplectic potentials $\theta_\alpha$ for $\omega$ in $V_\alpha$. The difference of two such forms $\theta_\alpha - \theta_\beta$ is a closed section of $\Lambda^1(V_\alpha \cap V_\beta)$. Again from the Poincaré Lemma we have the existence of $h_{\alpha\beta} \in W_\infty(V_\alpha \cap V_\beta)$ such that $(\theta_\alpha - \theta_\beta) = dh_{\alpha\beta}$ on $V_\alpha \cap V_\beta$. If the $h_{\alpha\beta}$ can be choosen so that $g_{\alpha\beta} = \exp(-\frac{i}{\hbar}h_{\alpha\beta})$ fulfils the cocycle condition $g_{\alpha\beta} g_{\beta\gamma} g_{\gamma\alpha} = 1$ we say that $\frac{1}{\hbar}\omega$ is integral.

In this case it is possible to construct a Hermitian line bundle B [18] over M with $g_{\alpha\beta}$ as transition functions and a connection $\nabla$ locally given by the expressions

$$(\nabla_X s)_\alpha = ds_\alpha(X) - \frac{i}{\hbar}\theta_\alpha(X)s_\alpha. \tag{6}$$

Here $s_\alpha$ are the the local representatives of the section s.

If $\frac{1}{\hbar}\omega$ is integral we define the prequantum bundle P over $(M,\omega)$ to be $P = B \otimes W^{1/2}$. Since we have constructed a local system for B with unitary transition functions, $P \otimes P^*$

$= W^1 \otimes C$. Since the transition functions constructed for P are in $W_\infty$ it is possible to define $P_\infty$; the set of $W_\infty$ sections of P.

For f such that $X_f$ exists it is possible to represent f as a differential operator on P by the local expressions

$$\rho(f)(s)_\alpha = -i\hbar X_f s_\alpha - \theta_\alpha(X_f)s_\alpha + fs_\alpha + \frac{i\hbar}{2}D_\alpha{}^*(X_f)s_\alpha. \tag{7}$$

**Proposition 4:** $\rho(f)$ *is a well defined differential operator on P.*

*Proof:* It is enough to show that $\rho(f)(s)_\beta = J_{\alpha\beta}{}^{1/2} g_{\alpha\beta} \rho(f)(s)_\alpha$. From $s_\beta = J_{\alpha\beta}{}^{1/2} g_{\alpha\beta} s_\alpha$ and lemma 2 we can write $\rho(f)(s)_\beta = -i\hbar X_f s_\beta - \theta_\beta(X_f)s_\beta + fs_\beta + \frac{i\hbar}{2}D_\beta{}^*(X_f)s_\beta$ as

$$J_{\alpha\beta}{}^{1/2} g_{\alpha\beta}\left[ -\frac{i\hbar}{2}D\ln J_{\alpha\beta}(X_f)s_\alpha - i\hbar D\ln g_{\alpha\beta}(X_f)s_\alpha - i\hbar X_f s_\alpha - \theta_\beta(X_f)s_\alpha + fs_\alpha \right.$$

$$\left. + \frac{i\hbar}{2}D\ln J_{\alpha\beta}(X_f) + s_\alpha\frac{i\hbar}{2}D_\alpha{}^*(X_f)s_\alpha\right].$$

Since $i\hbar D\ln g_{\alpha\beta} + \theta_\beta = \theta_\alpha$ it follows that this is equal to

$$J_{\alpha\beta}{}^{1/2} g_{\alpha\beta}\left[-i\hbar X_f s_\alpha - \theta_\alpha(X_f)s_\alpha + fs_\alpha - \frac{i\hbar}{2}D_\alpha{}^*(X_f)s_\alpha\right] = J_{\alpha\beta}{}^{1/2} g_{\alpha\beta} \rho(f)(s)_\alpha.$$

**Proposition 5:** $\rho(f)$ *is symmetric on* $P^\infty$, *i.e.*

$$<\rho(f)s,t> = <s,\rho(f)t>,$$

*where* $<s,t> = \int_M \mu_{st}*.$

*Proof:* Locally the measure $\mu_{\rho(f)st*} - \mu_{s\rho(f)t*}$ is given by $\varepsilon_\alpha$ times the function

$$[-i\hbar X_f s_\alpha - \theta_\alpha(X_f)s_\alpha + fs_\alpha + \frac{i\hbar}{2}D_\alpha{}^*(X_f)s_\alpha]t_\alpha{}^*$$

$$- s_\alpha[-i\hbar X_f t_\alpha - \theta_\alpha(X_f)t_\alpha + ft_\alpha + \frac{i\hbar}{2}D_\alpha{}^*(X_f)t_\alpha]^*$$

$$= -i\hbar X_f(s_\alpha t_\alpha{}^*) + i\hbar D_\alpha{}^*(X_f)(s_\alpha t_\alpha{}^*) = i\hbar s_\alpha t_\alpha{}^*[D\ln(s_\alpha t_\alpha{}^* - D^*\mu_{st*}(X_f)]).$$

This can be seen to be the local form of $i\hbar D^*\mu_{st*}(X_f)s_\alpha t_\alpha{}^*$. Hence

$$<\rho(f)s,t> - <s,\rho(f)t> = \int_M -i\hbar D^*{}_{st*}(X_f)\mu_{st*} = 0. \qquad (9)$$

**Proposition 6:** *When $X_f$ is complete there exists a one parameter group of isometries $\psi_t^*$ on $P_\infty$ such that*

$$\frac{d}{dt}\psi_t^* s = \frac{i}{\hbar}\rho(f)s. \qquad (10)$$

$\psi_t^*$ *is locally given by*

$$(\psi_t^* s)_\alpha(x) = exp(\frac{i}{\hbar}\int_0^t (\theta_\alpha(X_f) - f)dt')\, s_\alpha \circ \psi_t(x)\, J^\alpha{}_{\psi_t}{}^{1/2}(x). \qquad (11)$$

*if both $x$ and $\psi_t(x)$ belong to $U_\alpha$.*

*In (11) $\psi_t$ is the flow generated by $X_f$,*

$$J^\alpha{}_{\psi_t} = \frac{d\psi_t^* \varepsilon_\alpha}{\varepsilon_\alpha}$$

*and the integral $\int_0^t (\theta_\alpha(X_f) - f)dt'$ is the action integral along an integral curve.*

*Proof:* We will prove that $(\psi_t^* s)_\alpha$ as given in (11) transforms as a section of P. Start to look at how the factors of (11) transform.

$$\int_0^t \theta_\alpha(X_f) = \int_0^t \theta_\beta(X_f) + h_{\alpha\beta} \circ \psi_t - h_{\alpha\beta}$$

implies

$$exp(\frac{i}{\hbar}\int_0^t (\theta_\alpha(X_f) - f)dt') = exp(\frac{i}{\hbar}\int_0^t (\theta_\beta(X_f) - f)dt')\, g_{\alpha\beta} \circ \psi_t\, g_{\beta\alpha}.$$

Also

$$J^\alpha{}_{\psi_t} = \frac{d\psi_t^*(\varepsilon_\alpha)}{d\varepsilon_\alpha} = \frac{d\psi_t^*(\varepsilon_\alpha)}{d\psi_t^*(\varepsilon_\beta)}\frac{d\psi_t^*(\varepsilon_\beta)}{d\varepsilon_\beta}\frac{d\varepsilon_\beta}{d\varepsilon_\alpha} = J_{\alpha\beta} \circ \psi_t\, J_{\beta\alpha} J^\beta{}_{\psi_t}.$$

Hence

$$(\psi_t s^*)_\alpha(x) = exp(\frac{i}{\hbar}\int_0^t (\theta_\alpha(X_f) - f)dt')\, s_\alpha \circ \psi_t(x)\, J^\alpha{}_{\psi_t}{}^{1/2}(x) =$$

$$exp(\frac{i}{\hbar}\int_0^t (\theta_\beta(X_f) - f)dt') \; s_\alpha{}^\circ\psi_t(x) \; J^\beta{}_{\psi_t}{}^{1/2}(x) g_{\alpha\beta}{}^\circ\psi_t \; g_{\beta\alpha}(J_{\alpha\beta}{}^\circ\psi_t)^{1/2} \; J_{\beta\alpha}{}^{1/2} =$$

$$g_{\beta\alpha}{}^J{}_{\beta\alpha}{}^{1/2}exp(\frac{i}{\hbar}\int_0^t (\theta_\beta(X_f) - f)dt') \; s_\beta{}^\circ\psi_t(x) \; J^\beta{}_{\psi_t}{}^{1/2}(x),$$

which is the desired result. Any integral curve from x to $\psi_t(x)$ can be devided into pieces lying entirely within coordinate patches. $\psi_t$s is defined by composition of such transformations. For smooth enough sections one obtains (10) through differentiating (11) w.r.t. t.

REMARK. Since the flow $\gamma_t$ and $L_f$ are $W_\infty$ objects on M the action integral is not defined by pointwise integration over an integral curve. We expect the action $S_t$ to be a well defined object on M such that $\frac{d}{dt}S_t$ exists and is equal to $L_f$ a.e. $\varepsilon_\alpha$.

**Proposition 7:**

a)    $[\rho(f),\rho(g)] = -i\hbar\rho(\{f,g\})$                    (12)

b)    $\rho(c)s = cs$

for $f,g \in W_\infty$ and c a constant function.

*Proof*: b) is obvious and from the construction of the $\theta_\beta$s a) is true if and only if one can prove that $X_f D_\alpha{}^*(X_g) - X_g D_\alpha{}^*(X_f) = D_\alpha{}^*([X_f,X_g])$. This has been proved in [2] for vector fields X such that $DX \in HS(TM)$.

**4 Metaplectic structure on a Wiener manifold.**

First we consider some structures on the linear abstract Wiener space $(\mu,H,B)$.

**4.1 Particle structures.[17][18]**

A particle structure on H is a choice of complex structure J such that $\omega_0(x,y) = \omega_0(Jx,Jy)$ and

$$<x,y> = \omega_0(x,Jy) + i\omega_0(x,y)$$                    (13)

for all $x,y \in H$. Let $W_0 \subset H^C$ be the subspace of holomorphic directions w.r.t. J.

### 4.2 Fock-Cook quantization.[16][17][18]

On the symmetric tensor algebra $S(W_0)$ of $W_0$ there is an inner product defined for homogeneous elements by

$$<(e_1 \otimes ... \otimes e_n),(f_1 \otimes ... \otimes f_n)> = \Sigma_\pi <e_{\pi_1}f_1>...<e_{\pi_n}f_n> \qquad (14)$$

Let $S'(W_0)$ be the completion of $S(W_0)$ in this inner product. For $w \in W_0$ set

$$\rho(w)(e_1 \otimes ... \otimes e_n) = (w \otimes e_1 \otimes ... \otimes e_n) \qquad (15)$$

$$\rho(w^*)(e_1 \otimes ... \otimes e_n) = \Sigma_i <w,e_i>(e_1 \otimes .. \overset{i}{\wedge} .. \otimes e_n).$$

Then, for $x = \frac{1}{2}(w + w^*) \in H$,

$$\rho(x) = \frac{1}{2}[\rho(w) + \rho(w^*)]$$

is selfadjoint on $S'(W_0)$ and $\rho$ is a quantization of the linear observables on H.

### 4.3 Shale's representation.[16][18]

Let $Sp(H)$ be the group of continuous linear symmetries of $\omega_0$ and $U(H)$ the unitary group w.r.t. $<,>$. $U(H) \subset Sp(H)$ and $U(H)$ is exactly those $g \in Sp(H)$ for which $gW_0 = W_0$. $U(H)$ is a topological group w.r.t. the usual operator topology. Observe that the factors $u(g)$ and $|g|$ in the polar decomposition of a $g \in Sp(H)$ also belong to $Sp(H)$.

Let $Gl_2(H)$ be the set of positive elements of $Gl(H)$ having the form $(I + K)$ where $K \in HS(H)$. $Gl_2(H)$ is given the topology induced by the inner product $<K,L> = tr(KL^*)$ on $HS(H)$.

$Sp_r(H)$ is the set of $g \in Sp(H)$ such that $|g| \in Gl_2(H)$. $Sp_r(H) = U(H) \times Gl_2(H)$ as a set and $Sp_r(H)$ is a topological group with the factor topology of $U(H) \times Gl_2(H)$.

Shale has proved[18] that for $g \in Sp(H)$ there exists a unitary operator $U(g)$ on $S'(W_0)$ such that

$$U(g)\rho(x)U(g)^{-1} = \rho(gx) \text{ for all } x \in H \qquad (16)$$

if and only if $g \in Sp_r(H)$. $U(g)$ is determined up to a unitary factor. It is also proved in [18] that $g \to U(g)$ is a strongly continuous projective representation of $Sp_r(H)$ on $S'(W_0)$.

Let $Mp^C$ be the group of unitary operators $Z$ on $S'(W_0)$ such that there exists an element $g$ of $Sp(H)$ wich fulfils

$$Z\rho(x)Z^{-1} = \rho(gx) \text{ for all } x \in H. \tag{17}$$

We have the following exact sequence

$$1 \rightarrow U(1) \rightarrow Mp^C \rightarrow Sp_r(H) \rightarrow 1. \tag{18}$$

For a $g \in U(H)$ Shale's representation takes the form

$$U(g)(e_1 \otimes \ldots \otimes e_n) = (ge_1 \otimes \ldots \otimes ge_n).$$

Hence the restriction to $U(H)$ is a true representation which identifies $U(H)$ with a subgroup of $Mp^C$.

Having these objects it is possible to construct over $M$ the analogies of a Metaplectic frame bundle and halfforms.

## 4.4 Particle structures on M.

A particle structure on a Wiener manifold $(M,\omega)$ is a choice of Riemannian structure $<,>$ and complex structure $J$ on $TM$ such that $J$ is a particle structure for each tangent space. As before let $W_0$ be the isotropic subbundle of holomorphic directions w.r.t. $J$ in $TM^C$. The construction in [15] of an $Mp^C$ principal bundle over $M$ goes through also in the present context. The $Sp(H)$ reduction of $TM$ given by $\omega$ can further be reduced, using $W_0$, to a $U(H)$-structure which in turn can be extended both to a $Sp_r(H)$-structure $F(M)$ and an $Mp^C$-structure $E(M)$ in such a way that the diagram

$$\begin{array}{ccc} Mp^C & \rightarrow & E(M) \\ \downarrow & & \downarrow \\ Sp_r(H) & \rightarrow & F(M) \\ & & \downarrow \\ & & M \end{array} \tag{19}$$

commutes.

Let G(M) be the Hilbert space bundle over M associated with the action of $Mp^C$ on the Fock space. G(M) is isomorphic to the completion of the symmetric tensor product of the bundle $W_0$.

We thus have actions on G(M) both of $Mp^C$ through its natural action and of TM through the Fock-Cook representation $\rho(X)$ of the fibres of TM on the fibres of G(M). Also through Shale's representation we have a projective action of $Sp_r(H)$ on the fibres of G(M).

### 4.5 Lagrangian subspaces of $TM^C$.

A Lagrangian subspace of a fibre $TM_x^C$, $x \in M$, of $TM^C$ is a closed isotropic subspace W such that $W \oplus JW^* = TM_x^C$ . Let $D(M;W_0)$ be the set of all Lagrangian subspaces of $TM^C$ which can be reached by an element of $Sp_r(H)$ acting on $W_0$. $D(M;W_0)$ is a fibre bundle over M with fibre isomorphic to the restricted Siegel upper half plane, see [16]. $Sp_r(H)$ acts transitively on the fibre of $D(M;W_0)$ in the way shown in [16].

### 4.6 Halfforms on M.

Let $W \in D(M;W_0)$ be a Lagrangian subspace over $x \in M$. The complex line $\delta_W$ is defined as

$$\delta_W = \{q \in G(M)_x \; ; \; \rho(w)q = 0 \; for \; w \in W^*\}. \tag{20}$$

Then

$$\delta = \cup \; \delta_W, \, W \in D(M;W_0),$$

is a complex line bundle over $D(M;W_0)$ and $Mp^C$ acts on $\delta$ by the action induced from its action on G(M). Observe that $Z \in Mp^c$ takes $\delta_W$ to the fibre $\delta_{gW}$ where $g \in Sp_r(H)$ corresponds to Z through the relation (17). Through Shale's representation we have a projective action of $Sp_r(H)$ on $\delta$. See [16].

### 4.7 Action of canonical transformations on halfforms.

If $\phi$ is a canonical transformation on M the tangent mapping $\phi_*$ acts on the set of Lagrangian subspaces by $W \rightarrow \phi_*W$. Let $\phi$ be such that $D(M;W_0)$ is preserved under $\phi_*$. It can be shown that $\phi_*:TM_x \rightarrow TM_{\phi(x)}$ is of the form U(I+K) where $(I+K) \in Sp_r(H)$ acts on $TM_x$ and U is a unitary mapping $TM_x \rightarrow TM_{\phi(x)}$. U will preserve $W_0$. $\phi$ acts on G(M) letting (I+K) act on $G(M)_x$ through Shale's representation and

$$U(e_1 \otimes... \otimes e_n) = (Ue_1 \otimes... \otimes Ue_n) \; for \; (e_1 \otimes... \otimes e_n) \in G(M)_x.$$

Since $\delta_{Wx} \subset G(M)_x$, the action on $G(M)$ induces an action on $\delta$ such that

$$\phi_* \delta_{Wx} = \delta_{\phi_*(Wx)}.$$

### 4.8 Pairing.

For $W_1$ and $W_2 \in D(M, W_0)_x$ there is a pairing of $\delta_{W_1}$ and $\delta_{W_2}$ given by the inner product in $G(M)_x$.

REMARK. When H has finite dimension 2n there exists a character $\eta$ on $Mp^C$ whose kernel is Mp the double cover of Sp(H) This implies that E(M) locally reduces to an Mp-principal bundle. For any section W of $D(M;W_0)$ let det(W) be the line bundle of complex valued n-forms normal to W. Following [15] we have that

$$\delta_W \otimes \delta_W \otimes det(W) = P(\eta), \tag{21}$$

where $P(\eta)$ is the complex line bundle associated to $Mp^C$ through the character $\eta$.

When it is possible to make an Mp reduction of E(M) $\delta_W$ reduces to the symplectic spinors of Kostant[9] and $P(\eta)$ reduces to the trivial bundle. This means that in this case $\delta_W$ is isomorphic to a line bundle of $-\frac{1}{2}$-forms normal to W.

### 5 Polarizations.

A section W of $D(M;W_0)$ is a polarization if and only if W is spanned locally by hamiltonian vector fields. From our construction of $D(M;W_0)$ all polarizations are totally complex.

If $\phi$ is a canonical transformation on M and W is a polarization so is $\phi_*(W)$. It will be assumed that $W_0$ is a polarization.

To each section W of $D(M;W_0)$ belongs the line bundle $\delta_W = W^*\delta$ of halfforms normal to W. Here $W^*\delta$ denotes the pullback of $\delta$ via the section W. If $\phi$ preserves W then $\phi$ acts on $\delta_W$ by projecting the action $\phi_*$ on $\delta$.

Let $C_W$ be the set of all $f \in W_\infty$ such that $X_f$ exists and has a local flow $\gamma_t$ that preserves W. If $f \in C_W$ then the Lie derivative of a section $\tau$ of $\delta_W$ can be defined as

$$L_{X_f}\tau = \frac{d}{dt}\gamma_{-t*}(\tau). \tag{22}$$

$L_{X_f}$ is extended to complex hamiltonian vector fields by linearity.

As before we have the problem of existence of hamiltonian vector fields and integral curves.

## 6 Quantization of $(M, \omega)$.

The following is an outline of how we propose to do geometric quantization on a symplectic Wiener manifold.

### 6.1 Quantum bundle.

To each polarization belongs a quantum bundle $Q_W = P \otimes \delta_W$. For $f \in C_W$ let $\rho'(f)$ be the differential operator on $Q_W$ given by

$$\rho'(f)(s \otimes \tau) = (\rho(f)s \otimes \tau) + (s \otimes -i\hbar L_{X_f}\tau). \qquad (23)$$

It is clear that if $X_f$ is complete then $\rho'(f)$ generates a one-parameter group of transformations on $\Gamma(M, Q_W)$ locally given by $(\gamma_t{}^* \otimes \gamma_{-t*})$.

The pairings on $P$ and $\delta$ gives a pairing of $\Gamma(M, Q_W)$ and $\Gamma(M, Q_{W'})$.

$$<(s \otimes \tau), (s' \otimes \sigma)> = \int_M <\tau, \sigma> \mu_{(ss'^*)}, \qquad (24)$$

where $(s \otimes \tau) \in \Gamma(M, Q_W)$ and $(s' \otimes \sigma) \in \Gamma(M, Q_{W'})$ are local forms of the two factors in the pairing. Observe that all the $\delta_W$ are trivial line bundles so that any section of $Q_W$ can be written as a global tensor product.

Let $\nabla'_X$ be the differential operator on $P$ which has the local form

$$(\nabla'_X s)_\alpha = X s_\alpha - \frac{i}{\hbar}\theta_\alpha(X)s_\alpha - \frac{1}{2}D_\alpha{}^*(X)s_\alpha. \qquad (25)$$

A section of $Q_W$ is polarized if it can be written locally as $(s \otimes \tau)$ where $s$ and $\tau$ obey the differential constraints

$$\nabla'_{X_f} s = 0 \; and \qquad (26)$$

$$L_{X_f}\tau = 0 \; for \; all \; X_f \in W. \qquad (27)$$

There is a canonical section of $\delta_{W_0}$ given by the vacuum state in $G(M)_x$. This section is polarized but for a general $\delta_W$ it is not clear wheather polarized sections exist or not.

The following propositions should follow from the definitions.

**Proposition 1.** *If $f \in C_W$ $\rho'(f)$ preserves the set of polarized sections of $Q_W$.*

**Proposition 2.** *Let $f \in W_\infty$, $\gamma_t$ the flow generated by $X_f$, $W = \gamma_{-t*}(W_0)$ and $(s \otimes \tau)$ a polarized section w.r.t. $W_0$ then $(\gamma_t^* s \otimes \gamma_{-t*}\tau)$ is polarized w.r.t. $W$.*

**Proposition 3.** *If $v$ and $v'$ are polarized sections of $Q_W$ and $Q_{W'}$ respectivelly and $f$ preserve both $W$ and $W'$ then $<\rho'(f)v,v'> = <v,\rho'(f)v'>$ when both sides are defined.*

**Proposition 4.**

$$[\rho'(f),\rho'(g)] = -i\hbar\rho'(\{f,g\}).$$

$$\rho'(c)v = cv$$

*for $f$ and $g \in C_W$, $c$ a constant function and $v$ a section of $Q_W$.*

Assuming these propositions we can make the following constructions.

## 6.2 Quantization.

Let $\mathcal{H}_W$ be the Hilbert space obtained by completing the set of polarized sections in $\Gamma(M,Q_W)$ w.r.t. the inner product obtained by restricting the pairing (24) to $\Gamma(M,Q_W)$. By proposition 3 for $f \in C_W$ $\rho'(f)$ is a symmetric operator on $\mathcal{H}_w$.

$\rho'(f)$ is a representation of the subalgebra $C_W$ of $W_\infty(M)$ on $\mathcal{H}_W$. When $W = W_0$, $\rho'(f)$ is the restriction of prequantization to $\mathcal{H}_W$ since $L_{X_f}\tau = 0$ for all $X_f$ preserving $W_0$. This corresponds to the skipping of divergent trace terms, e.g. $\sum_1^\infty \frac{1}{2}$ for a harmonic oscillator, which thus is automaticaly incorporated in this quantization procedure.

## 6.3 BKS[3][17][18]

For a general $f \in W_\infty(M)$ such that $X_f$ is complete, using proposition 2, $(\gamma_t^* s \otimes \gamma_{-t*}\tau)$ is a polarized section of $Q_{\gamma_{t*}(W)}$ . By proposition 3, the pairing (24) intertwines the representations of $C_W$ and $C_{\gamma_{t*}(W)}$ on $Q_W$ and $Q_{\gamma_{-t*}(W)}$ respectively. It is then natural to use the BKS-construction of $\rho'(f)$ in this case.

When $f \in W_\infty(M)$ is such that $X_f$ is complete $\rho'(f)$ is defined through the equation

$$<\rho'(f)(s \otimes \tau),(s' \otimes \sigma)> = -i\hbar\frac{d}{dt}<(\gamma_t^* s \otimes \gamma_{-t*}\tau),(s' \otimes \sigma)> /_{t=0} \quad (28)$$

when the right-hand side is defined.

This will in general extend $\rho'$ to a larger subset of $W_\infty(M)$ but it will also introduce anomalies. Shale´s results [18] shows that anomalies can not be avoided in general.

## 7 Linear phase space.

Here we apply the proposed machinery to an abstract linear Wiener space and obtain the usual quantization of a linear system.

Let $(\mu, H, B)$ be an abstract Wiener space with a strong symplectic form $\omega$ on H. The chart $(U_0, \varphi_0) = (B, \mathrm{id}_B)$ gives B the structure of a Wiener manifold and the background measure $\varepsilon_0 = \mu$. For any $g \in Gl_2(H)$ $(B, g)$ is a new chart of B. The transition function $g \circ \varphi_0^{-1} = g$ fulfils the condition $g \circ \varphi_0^{-1} - I \in H\text{-}C^\infty(B \to H)$ and

$$J_{g0} = J_g = \chi(I+K)exp(-"<u,Ku> - tr(DK)" - \frac{1}{2}<Ku,Ku>),$$

where $g = I + K$ and $\chi$ is the Carleman-Fredholm determinant. The measure $\varepsilon_g$ given by $d\varepsilon_g = J_{g0}d\mu$ will be the Guassian measure associated with the inner product $<h,h'>_g = <gh,gh'>$ on H.

Choose a complex structure $J_0$ on H such that $<,>_H = \frac{2}{\hbar}\omega(,J_0)$ and let $\theta$ be the symplectic potential given by $\theta(X) = \omega(u,X)$. $W_0$ will denote the holomorphic directions in H w.r.t. $J_0$.

If $g \in Sp(H) \cap Gl_2(H)$ then

$$<h,h'>_g = <gh,gh'> = \frac{2}{\hbar}\omega(gh,J_0gh') = \frac{2}{\hbar}\omega(h,g^{-1}J_0gh') = \frac{2}{\hbar}\omega(h,J_Wh')$$

where $J_W$ is the complex structure corresponding to the Lagrangian subspace $W = g^{-1}W_0$. In this case $J_g = exp(-\frac{1}{\hbar}\omega(u,(J_W - J_0)u))$.

## 7.1 Prequantum bundle.

In what follows we will make use of the following proposition.

**Proposition:** *For f such that $D^*(X_f)$ is well defined $tr(DX_f) = 0$ and $D_o^*(X_f) = "<u,X_f>"$. Here $D_0$ is the dual of D w.r.t. the measure $\varepsilon_0$.*

*Proof:* This is certainly true for linear observables and using finite dimensional symplectic subspaces in taking the limit defining $D^*(X_f)$ one sees that the trace term vanishes and "$<u,X_f>$" exists as a limit in $L_2(\mu)$.

The prequantum bundle P is trivial and using the coordinate patch (B,g) one can identify $P_2$ with $L_2(\varepsilon_g, B)$. The coordinate change g corresponds to a change of basis in P given by

$$J_{g0}^{1/2} = exp(-\frac{1}{2\hbar}\omega(u,(J_W - J_{W_0}u), \quad i.e. \quad \Psi_0 = exp(-\frac{1}{2\hbar}\omega(u,(J_W - J_{W_0})u)\Psi_g. \qquad (28')$$

$$W = g^{-1}W_0.$$

In the coordinate system (B,g) the prequantum operator corresponding to an observable f will have the form

$$\rho(f) = -i\hbar X_f - \frac{1}{2}\omega(u,X_f) + f + \frac{i\hbar}{2}D_g^*(X_f). \qquad (29)$$

Observe that since linear symplectic coordinate changes have been used, $\theta(X) = \omega(u,X)$ has the same form in all trivializations.

Also $D_g^*(X) = D_0^* + DlnJ_{g0} = D_0^*(X) + \frac{2}{\hbar}\omega(u,(J_w - J_0)X)$. This means that $\rho(f)$ takes the form

$$\rho(f) = -i\hbar X_f - \omega(u,(I - iJ_W)X_f) + f \qquad (30)$$

in the coordinate system (B,g), for $g \in Sp(H) \cap Gl_2(H)$. The prequantum operator corresponding to the linear observable $l_X$, given by $l_X(u) = 2\omega(u,X)$ for X a constant vector field, is

$$\rho(l_X) = -i\hbar X + \omega(u,(I + iJ_W)X) \qquad (31)$$

When $X_f$ has complete flow $\gamma_t$*s takes the form $exp(-\frac{i}{\hbar}\int_0^t L_f dt') \Psi \circ \gamma_t(x) J_{\gamma_t}^{1/2})$.

## 7.2  Polarizations.

Let $W = g^{-1}W_0$. In the trivialization corresponding to (B,g) the differential constraints

$$\nabla'_X s = 0 \text{ for all } X \in W^*, \qquad (32)$$

describing the polarized sections w.r.t. W takes the form

$$X\Psi_g = 0 \text{ for all } X \in W^*. \qquad (33)$$

The quantum Hilbert space $H_W$ of polarized sections w.r.t. W is thus the set of holomorphic functions in $L_2(\varepsilon_g, B)$. The linear observables act on a function $\Psi \in H_W$ in the following way

$$\rho(l_X)\Psi = \omega(u,(I + iJ_W)X)\Psi = l_X\Psi \ for \ X \in W^* \ and \qquad (34)$$

$$\rho(l_X)\Psi = -i\hbar X\Psi \qquad\qquad for \ X \in W.$$

## 7.3 Metaplectic structure on B.

Each Lagrangian subspace W of $H^C$ corresponds to a so called constant polarization in the natural way. $D(W_0,B)$ is a trivial line bundle over B isomorphic to $B \times D$, where D is the restricted Siegal upper half plane consisting of $Z \in HS(H)$ such that

$$ZJ = -JZ \qquad (35)$$

$$<Zx,y> = <Zy,x>$$

$$<Zx,x> < <x,x>$$

There is a 1-1 correspondence between D and the Lagrangian subspaces W of $H^C$ such that $J_w - J_{w_0} \in HS(H)$. This correspondence maps $W_0$ to 0.

$\delta$ is also a trivial line bundle over $D(W_0,B)$. $\delta = B \times D \times C$. Let $\varepsilon_{W,x}$ be the section of $\delta$ such that $\varepsilon_{w_0,x}$ is the Dirac vacuum and

$$<\varepsilon_{W,x},\varepsilon_{W',x}> = det(I - Z_1{}^*Z_1)^{-1/4}det(I - Z_2{}^*Z_2)^{-1/4}det(I - Z_1{}^*Z_2)^{1/2} \qquad (36)$$

Since the linear observables has a trivial action on the tangent spaces and W is a constant polarization, $x \to \varepsilon_{W,x}$ is a polarized section of $\delta_W$.

For $g \in Sp_r(H)$

$$g_*\varepsilon_W = arg(det(I + a^{-1}bZ))^{1/2}\varepsilon_{gW}.$$

See [16] for notation.

## 7.4 Quantization.

Choose some constant polarization corresponding to a Lagrangian subspace W. Then $Q_W = \delta_W \times P$ and a polarized section of $Q_W$ has the form $\Psi_W \otimes \varepsilon_W$, with $\Psi_W$ holomorphic w.r.t. W, in the trivialization given by (B,g). As before g is a symplectic mapping such that $g^{-1}W_0 = W$. Since the linear observables act trivially on $\varepsilon_W$ the quantization of the linear observables looks like

$$\rho'(l_X)\Psi_W \otimes \varepsilon_W = \omega(u,(I + iJ_W)X)\Psi_W \otimes \varepsilon_W = l_X\Psi_W \otimes \varepsilon_W \ for \ X \in W^* \quad (37)$$

$$\rho'(l_X)\,\Psi_W \otimes \varepsilon_W = (-i\hbar X\,\Psi_W) \otimes \varepsilon_W \ \ for\, X \in W.$$

Since $<\varepsilon_{W,x},\varepsilon_{W,x}> = 1$ the inner product of two polarized sections s and t takes the form

$$<s,t> = \int_M \Psi_W \, \Psi''^*_W \, d\varepsilon_g. \tag{38}$$

This is the usual form of the Bargman quantization.

## 7.5 Pairing.

Next we want to consider the BKS pairing between two sections s and t polarized w.r.t. W and W' respectively. We know that s and t have the forms $\Psi_W \otimes \varepsilon_W$ and $\Psi_{W'} \otimes \varepsilon_{W'}$, where $\Psi_W$ and $\Psi_{W'}$ are holomorphic w.r.t. W and W' respectively, in the trivializations corresponding to (B,g) and (B,g') respectively. However, to express the pairing we need the forms of s and t in a common trivialization of P. From (28´) it follows that in (B,id)

$$s = \ exp(-\frac{1}{2\hbar}\omega(u,(J_W - J_{W_0})u)\,\Psi_W \otimes \varepsilon_W \tag{39}$$

$$t = exp(-\frac{1}{2\hbar}\omega(u,(J_{W'} - J_{W_0})u)\,\Psi_{W'} \otimes \varepsilon_{W'}$$

$$<s,t> = <\varepsilon_W,\varepsilon_{W'}> \int_M \Psi_W \Psi^*_{W'} \, exp(-\frac{1}{2\hbar}\omega(u,(J_W + J_{W'})u) + \frac{1}{\hbar}\omega(u,J_{W_0}u))d\mu$$

One easily convices oneself that

$$exp(-\frac{1}{2\hbar}\omega(u,(J_W + J_{W'})u) + \frac{1}{\hbar}\omega(u,J_{W_0}u)) = \frac{d\mu_G}{d\mu},$$

where $\mu_G$ is the Gaussian measure corresponding to the inner product $G(u,u) = \frac{1}{\hbar}\omega(u,(J_W + J_{W'})u)$ on H. Observe that $<\varepsilon_W,\varepsilon_{W'}>$ has the form given in [16]. This is the standard pairing leading to the Bogoliubov transformation between the two Fock spaces associated with the Lagrangian subspaces W and W', see [24].

## 7.6 Action of $Sp_r$.

We will consider the two cases, $g \in U(H)$ and $g \in Sp(H) \cap Gl_2(H)$. In general a $g \in Sp_r(H)$ will not have a hamiltonian action on B since the would be hamiltonian function do not extend to a regular enough function on B. However $\theta$ is constructed so that $L_f = 0$ for quadratic functions f. Also since a $g \in U(H)$ preserve $\mu$ we will put $J_g = 1$ in this case even though since $g \notin Gl_2(H)$ $J_g$ is not well defined.

Let $s = \Psi_{W_0} \otimes \varepsilon_{W_0}$ be a polarized section of $Q_{W_0}$ and $g \in U(H)$. From the form of the actions of $Sp_r$ on sections of P and $\delta$ in paragraphs 7.1 and 7.4 the lifting of a unitary g to an action on sections of $Q_{W_0}$ has the form

$$\gamma_g^*(\Psi_{W_0} \otimes \varepsilon_{W_0}) = J_g^{1/2} \Psi_{W_0}{}^{\circ}g \otimes \varepsilon_{g^{-1}W_0} = \Psi_{W_0}{}^{\circ}g \otimes \varepsilon_{W_0}. \qquad (40)$$

For $g \in Sp(H) \cap Gl_2(H)$ $J_g$ is well defined. g thus directely lifts to actions on P and $\delta$

$$\gamma_g^*(\Psi_{W_0} \otimes \varepsilon_{W_0}) = J_g^{1/2} \Psi_{W_0}{}^{\circ}g \otimes \varepsilon_{g^{-1}W_0} = exp(-\frac{1}{2\hbar}\omega(u,(J_W - J_{W_0})u)\Psi_{W_0}{}^{\circ}g \otimes \varepsilon_W. \quad (41)$$

W is the Lagrangian subspace $g^{-1}W$. $\Psi_{W_0}{}^{\circ}g$ is holomorphic w.r.t. W so that

$$exp(-\frac{1}{2\hbar}\omega(u,(J_W - J_{W_0})u)\Psi_{W_0}{}^{\circ}g$$

is the form of a polarized section w.r.t. W in the trivialization (B,id). If one transforms $\gamma_g^*(s)$ to (B,g) one obtains $\Psi_{W_0}{}^{\circ}g \otimes \varepsilon_W$ which conforms with the form used in [23].

## 7.7 Infinitesimal Pairing.

For a general observable $\rho'(f)$ if it exists is defined by the equation

$$<\rho'(f)s,s'> = -i\hbar\frac{d}{dt}<\gamma_t^*s,s'>/_{t=0}, \qquad (42)$$

where s and s' are polarized sections of $Q_{W_0}$. Let $I(s,s') = <\gamma_t^*s,s'>$. If one writes $s = \Psi_{W_0} \otimes \varepsilon_{W_0}$ and $s' = \Psi'_{W_0} \otimes \varepsilon_{W_0}$ then $\gamma_t^*s$ takes the form

$$\gamma_t^*s = J_{\gamma_t}^{1/2} \Psi_{W_0}{}^{\circ}\gamma_t \otimes U(\gamma_{t*})\varepsilon_{W_0} \qquad (43)$$

and

$$I(s,s') = \int_M <U(\gamma_{t*})\varepsilon_{W_0},\varepsilon_{W_0}>J_{\gamma_t}^{1/2}exp(-\frac{i}{\hbar}S_t)\Psi_{W_0}{}^{\circ}\gamma_t\Psi'_{W_0}d\mu, \qquad (44)$$

where $S_t = \int_0^t L_t dt'$. Differentiating under the integral sign and using the identities

$$<U(\gamma_{t*})\varepsilon_{W_0},\varepsilon_{W_0}> = 1 + O(t^2) \qquad (45)$$

$$\frac{d}{dt}(S_t) = L_f \qquad (46)$$

one obtains

$$\frac{d}{dt}I(s,s') = \int_M (X_f - \frac{i}{\hbar}L_f - \frac{1}{2}D_0^*(X_f))\Psi_{W_0}\Psi'^*_{W_0}d\mu. \qquad (47)$$

Let $X_f^+$ and $X_f^-$ be the holomorphic and antiholomorphic parts of $X_f$ respectively. Using the fact that $\Psi_{W_0}$ and $\Psi'^*_{W_0}$ are holomorphic w.r.t. $J_0$ and the equation $X_f = J_0Df$ one performs a partial integration

$$\frac{d}{dt}I(s,s') = \int_M (-\frac{i}{\hbar}L_f + \frac{1}{2}D_0^*(X_f^+ - X_{f'})) \Psi_{W_0} \Psi''^* w_0 d\mu$$

$$= \int_M (-\frac{i}{\hbar}L_f + \frac{1}{2}D_0^*(iDf)) \Psi_{W_0} \Psi''^* w_0 d\mu = \int_M (-\frac{i}{\hbar}L_f + \frac{i}{2}\Delta_{OU}f) \Psi_{W_0} \Psi''^* w_0 d\mu) \quad (48)$$

Here $\Delta_{OU}f = D_0^*Df$ is the Ornstein-Uhlenbeck operator on $(\mu, H, B)$. If it is possible to construct reproducing kernels for the Bargman spaces $H_W$ this expression will give us $\rho'(f)$ in analogy with the constructions in [22] and [25].

REMARK. Observe that since $<,>$ contains a factor $\frac{1}{\hbar}$ so does the Ornstein-Uhlenbeck operator.

## 8 References

[1] Andersson, L., Functional integration and geometric quantization, Stochastic analysis, path integration and dynamics ed. K.D. Elworthy and J-C Zambrini, Longman Sientific and Technical, pp. 22-44, 1989.

[2] Andersson, L., Prequantization of infinite dimensional dynamical systems, J. Func. Anal., vol. 75, pp. 58-91, 1985.

[3] Ashtekar A. and Magnon A., A geometrical approach to external potential problems in quantum field theory, Gen. Rel. Grav., vol 12, pp. 205-223, 1985.

[4] Blattner, R. J., The meta linear geometry of non real polarizations, Springer Lect. notes in Math. vol. 570 Differential geometric methods in math. phys. ed.Bleuler, pp. 11-45, 1975.

[5] Bowick, M. J. and Rajeev, S.G., The holomorphic geometry of closed bosonic string theory and Diff(S¹)/S¹, Nuclear Physics B293, pp. 348-384, 1987.

[6] Cruzeiro, A.B., Unicité de solution d'equations differentialles sur l'espace de Wiener, J. Func. Analysis, vol. 58, pp.335-347, 1984.

[7] Eells, J. and Elworthy, K.D., Wiener integration on certain Manifolds, C.I.M.E., IV ´Some Problems in Nonlinear Anaysis´,pp. 67-94, 1971.

[8] Gross, L., Abstract Wiener spaces, Proc. 5th Berkleley Symp., vol. 2, pp.97-109, 1977.

[9] Kostant, B., Symplectic spinors, Symposia Mathematica, 14, pp. 139-152, Academic Press,1974.

[10]   Kusuoka, S., On the foundations of Wiener-Riemannian manifolds, Stochastic
       analysis, path integration and dynamics ed. K.D. Elworthy and J-C Zambrini,
       Longman Sientific and Technical, pp. 130-164, 1989.

[11]   Malliavin, P., Implicit functions in finite corank on the Wiener space, Stocastic
       Anal, Proc. of Taniguchi Intern. Symp. ed. by K. Ito, pp. 369-386, 1984.

[12]   Meyer, P. A., Quelques resultats analytiques sur le semi groupe d'Ornstein-
       Uhlenbeck en dimension infinie, Springer lecture notes in Control and Inf. Sci.
       **49**, pp. 201-214,1983.

[13]   Piech, M. A., The exterior algebra for Wiemann manifolds, J Func. Anal., vol.
       **28**, pp. 279-308, 1978.

[14]   Ramer, R., On Nonlinear Transformations of Gaussian measures, J Func. Anal.,
       vol. **15**, pp. 166-187, 1974.

[15]   Ramnsley, J.H. and Robinson, P.L., Mp$^C$-structures and geometric quantization,
       preprint Univ. of Warick, 1984.

[16]   Segal, G., Unitary representations of some infinite dimensional groups, C.M.P.,
       vol. **80**, pp. 301-342, 1981.

[17]   Segal, I. E., Tensor algebras over Hilbert spaces I, Trans. Amer. MAth. Soc.
       vol. **81**, pp.106-134, 1956.

[18]   Shale, D., Linear Symmetries of free boson fields, Trans. Am. Soc., vol. **103**,
       pp. 149-167,1962.

[19]   Shigekawa, I., De Rahm-Hodge-Kodaira's decomposition on an abstract Wiener
       space, J. Math. Kyoto Univ. vol 26, pp.191-202, 1986.

[20]   Sternberg, S., The pairing methods and bosonic anomalies, Diff. Geom. methods
       in Theo. Phys., Kluwer Academic Press, 1988.

[21]   Sugita, H., Sobolev spaces of Wiener functionals and Malliavin's calculus, J.
       Math. Kyoto Univ., vol. **25**, pp. 31-48,1985.

[22]   Tuynman, T. M., Studies in geometric quantization, Thesis, Univ. Amsterdam,
       1988.

[23]  Woodhouse, N. M. J., Geometric quantization and the Bogoliubov
      Transformation, Proc. R. Soc. Lond., vol A378, pp. 119-139, 1981., 1981.

[24]  Woodhouse, N. M. J., Geometric quantization, Clarendon Press, Oxford, 1980.

[25]  Woodhouse, N. M. J., The BKS method in Fock space, Preprint Woodham
      Collage Oxford.

[26]  Isham C.J. and A.C. Kakas, A group theoretical approach to the quantization of
      gravity: I and II, Class. Quantum Grav. 1, pp. 621-650, 1984.

# Heat Kernels on Lie Groups

Maria Teresa Arede

**Abstract:**

Using the Elworthy–Truman "elementary" formula we obtain exact and explicit formulae for the heat kernel on Lie groups and their dual symmetric spaces. We analyze also the case of nilpotent Lie groups and by means of a faithful representation we obtain for the heat kernel, associated with the Laplace–Beltrami, a recursion formula on the dimension of the representation.

## 0. Introduction

In this paper, we discuss the heat equation on manifolds in particular on Lie groups. A motivation for this can be a better understanding of the relations between classical and quantum systems. Such relations can be looked upon in the following way.

Consider the Schrödinger or heat equation (without potential) on a Riemannian manifold $M$. We have resp.

$$(i\frac{\partial}{\partial t} + \frac{\hbar}{2}\Delta)\, u(t,x) = 0$$

$$(\frac{\partial}{\partial t} - \frac{\hbar}{2}\Delta)\, u(t,x) = 0$$

where $\Delta$ is the Laplace–Beltrami operator. The evolution group, resp. semigroup, is given by

$$u(t,x) = e^{i\hbar/t\Delta/2} u_0(x)$$

resp.

$$u(t,x) = e^{\hbar t\Delta/2} u_0(x) \tag{1}$$

where $u_0$ is the initial condition, $u_0(x) = u(0,x)$.

For expressions (1) we can then say that the asymptotic behavior for $\hbar/\downarrow 0$ (semiclassical behavior) is equivalent to the asymptotic behavior for $t \downarrow 0$ (small time asymptotics).

For a Riemannian manifold it is well known that (see, e.g., [12] and [3]) the fundamental solution of the heat equation, denoted by $p(t,x,y)$, has for $y \notin Cut(x)$, $(Cut(x)$ standing for the cut locus of $x)$, the following asymptotic behavior

$$p(t, x, y) \sim (2\pi t)^{-n/2}\theta^{-1/2}(x, y)e^{-d^2(x,y)/2t} \tag{2}$$

for $t \downarrow 0$, where

$$\theta(x, y) = [\det g_{ij}(\exp_x t\dot{\gamma}(0))]^{-1/2}$$

is the "Ruse invariant" with $(g_{ij})$ denoting the matrix representating the Riemannian metric in local coordinates in a neighborhood of $x$ and where $\gamma$ is the unique geodesic between $x$ and $y$ with $\gamma(0) = x$, $\gamma(t) = y$ and $\dot{\gamma}(0)$ is the tangent vector at 0.

We notice that (see, e.g., [6])

$$[\det g_{ij}(\exp_x t\dot{\gamma}(0))]^{-1/2} = \det T_{t\dot{\gamma}(0)} \exp_x$$

where $T_{t\dot{\gamma}(0)}$ denotes the tangent mapping of $\exp_x$ at $t\dot{\gamma}(0)$.

We would like then to know, and this is the main idea of this work, for which classes of Riemannian manifolds, especially Lie groups, is the heat kernel given in an exact way – that is for all times – by an expression of type (2) or by a sum, over all geodesics between $x$ and $y$, of terms of that form.

Besides the interest of exact formulae in themselves, such results enable us to better understand the relations between $p(t, x, y)$ and the geometric aspects of the manifold, e.g., the lengths of geodesics and the curvature of the manifold.

The content of this paper is as follows. In the first section we present the main tools which will be used in Section 2 and we discuss also some general applications. In Section 2 we obtain explicit formulae for the heat kernel on compact Lie groups and on their dual symmetric spaces. In the third section we study the case of nilpotent Lie groups obtaining by a different technique, recursive expressions for the heat semigroup as well as for its kernel.

We conclude with some remarks and possible developments.

## 1. The Elworthy–Truman formula and applications

Let us consider a connected, complete Riemannian manifold $M$, such that it possesses a pole, that is, there exists a point $y_0 \in M$ for which the exponential map

$$\exp_{y_0} : T_{y_0}M \to M$$

is a diffeomorphism.

Like before we denote

$$\theta_{y_0}(x) = |\det T_X \exp_{y_0}|$$

for $x \in M$ and $X = \exp_{y_0}^{-1} x$, the determinant being taken with respect to orthonormal bases in $T_{y_0}M$ and $T_x M$. As $y_0$ is a pole, we always have $\theta_{y_0}(x) \neq 0$.

Suppose now that $\| \log \theta_{y_0}( ) \|$ is a bounded function on $M$ and that $\Delta \log \theta_{y_0}( )$ is bounded above. We have the following:

**Theorem 1.1.** ([8], [9]) *Under the above conditions and if $V : M \to \mathbf{R}$ is a locally Hölder continuous function on $M$, then the fundamental solution, $p(t, x, y)$, of the equation*

$$\left( \frac{\partial}{\partial t} - \frac{1}{2}\Delta + V \right) u(t, x) = 0 \tag{3}$$

*with $\Delta$ being the Laplace–Beltrami operator on $M$, is given by*

$$p(t, x, y_0) = (2\pi t)^{-n/2} \theta_{y_0}^{-1/2}(x) e^{-d^2(x, y_0)/2t}$$
$$\times \mathbf{E} \left\{ \exp \left( \frac{1}{2} \int_0^t \theta_{y_0}^{1/2}(x_s) \Delta_{y_0}^{-1/2}(x_s) + V(x_s) ds \right) \right\}$$

*where $d(x, y_0)$ is the geodesic distance between $x$ and $y_0$ and the expectation is taken with respect to the Riemannian Brownian bridge $(x_s)_{0 \leq s \leq t}$ associated to the drift*

$$A_s = -\nabla_x \frac{d^2(x, y_0)}{2(t - s)} - \frac{1}{2}\nabla_x \log \theta_{y_0}(x)$$

*and conditioned to start at $x_0 = x$ and to stop at $x_t = y_0$.*

*Moreover the radial component of $(x_s)$, denoted by $r_s = d(x_s, y)$, has the same probability distribution as the radial component of the canonical Brownian bridge on $\mathbf{R}^n$ between $v \in \mathbf{R}^n$ and the origin, for $v$ such that $\|v\| = d(x_0, y_0)$.*

This result (as well as an extension of it, see, e.g., [9] and [2]) can be applied in several cases, namely: simply harmonic spaces (for which $\theta_{y_0}(x) = 1$), spaces of constant negative sectional curvature (like hyperbolic spaces and Clifford–Klein spaces forms) and also symmetric spaces of noncompact type, duals of compact Lie groups. For details see [2].

As it can be seen in [8], this result is also very useful in the study of the asymptotic behavior of $p(t, x, y)$ as $t \downarrow 0$.

Another important result in what follows is:

**Theorem 1.2.** ([14]) *Let $M$ be a connected complete Riemannian manifold and $y \in M$. Suppose that $Cut(y)$ has codimension greater than one. Then the fundamental solution of (3), $p(t, x, y)$, is, for $x \in M -$*

$Cut(y)$, given by

$$p(t, x, y) = (2\pi t)^{-n/2} \theta_y^{-1/2}(x) e^{-d^2(x,y)/2t}$$

$$\times \mathbf{E}_x \left\{ \chi_{\tau > t} \exp \left( \frac{1}{2} \int_0^t \theta_y^{1/2}(x_s) \Delta \theta_y^{-1/2}(x_s) + V(x_s) ds \right) \right\}$$

where $\chi_{\tau > t}$ is the indicator function of the set $\{\omega | \tau(\omega) > t\}$ for $\tau$ the first exit time of $(x_s)_{0 \leq s \leq t}$ from $M - Cut(y)$.

This result can be used in a more general context than the first one, as we will see in the next section.

## 2. Heat kernel on compact Lie groups
## and on its dual symmetric spaces

### 2.1. The heat kernel on compact Lie groups

In this section, $G$ denotes a compact Lie group and $\mathbf{G}$ its Lie algebra. The canonical exponential map is

$$\exp : \mathbf{G} \to G$$

where $\exp X = \gamma(1)$ for $X \in \mathbf{G}$ and $\gamma : \mathbf{R} \to G$ is a homomorphism such that $\gamma(0) = e$ and $\dot{\gamma}(0) = X$, $e$ denoting the identity of $G$.

In the case of a compact Lie group $G$ we have a canonical biinvariant Riemannian structure, namely the one induced by the Killing form $\mathbf{G}$ (see, e.g., [12]). As a consequence, we have that the mapping exp above coincides with $\exp_e$ defined using only the manifold structure of $G$. Equivalently we can say that the geodesics in $G$ coincide with the one-parameter subgroups of $G$.

We can now apply Theorem 1.2 to be the open set $G - Cut(e)$.

Indeed we remark that $Cut(e)$ is given by the first conjugate points of $G$, more precisely

$$Cut(e) = \{\exp(Ad(g)H) | H \in \mathbf{T} : i\alpha(H)$$
$$= \pm 2\pi \quad \text{for some } \alpha > 0 \text{ and for all } g \in G\}$$

where $Ad$ denotes the adjoint representation of $G$, $\mathbf{T}$ is the Cartan subalgebra associated to the maximal torus $T$ and $\alpha$ are the roots of $G$ with respect to $T(\alpha > 0$ being the positive roots); see, e.g., [7], [2].

On the other hand

$$\dim \ \text{conj}\,(G) = \dim G - 3$$

where $\text{conj}(G)$ are the points in $G$ conjugated of the identity $e$. So $Cut(e)$ has codimension greater than two (or, in another way, it has capacity zero with respect to the Brownian motion on $G$). We can then write the fundamental solution of

$$\left(\frac{\partial}{\partial t} - \frac{1}{2}\Delta\right) u(t, x) = 0$$

in $G$, is given by

$$p(t, x, e) = (2\pi t)^{-n/2}\theta_e^{-1/2}(x)e^{-d^2(x,e)/2t}$$

$$\times \mathbf{E}_x\left\{\chi_{\tau > t} \exp\left(\frac{1}{2}\int_0^t \theta_e^{1/2}(x_s)\Delta\theta_e^{-1/2}(x_s)ds\right)\right\}$$

where $x \in G - Cut(e)$ and all expressions have the same meaning as in Theorem 1.2.

We remark now that exp is a surjective map and then every $x \in G$ is of the form $x = \exp X$ for some $X \in G$. On the other hand, we have $x = uhu^{-1}$ for some $h \in T$ and $u \in G$. Then

$$x = uhu^{-1} = u\exp Hu^{-1} = \exp Adu(H)$$

where $H \in T$.

The fundamental solution, $p(t, x, e)$, being invariant under right and left translations can be written as

$$p(t, x, e) = p(t, h, e) = p(t, \exp H, e)$$

where $H \in T$.

Let us now consider the following two lemmas.

**Lemma 2.1.** *The function*

$$\theta_e(x) = |\det T_x \exp|$$

*is a "class function," that is, it depends only on the elements of* **T** *(or on the elements of T) and is given by*

$$\theta_e(x) = \theta_e(\exp H) = \prod_{\alpha > 0}\left(\frac{2\operatorname{sen} i\alpha(H)/2}{i\alpha(H)}\right)^2$$

*where* $\alpha > 0$ *are the positive roots of $G$ with respect to $T$.*

**Proof.** See [2].

Before the second lemma, let us point out that the action of the Laplace–Beltrami operator, $\Delta$, over functions defined on **T** and invariant by the Weyl group is given by its radial part, that is (see, e.g., [5])

$$\Delta_{\text{rad}} = \frac{1}{j}\left(\sum_{i=1}^{1}\frac{\partial^2}{\partial h_i^2} + <\theta, \theta>\right) j$$

where

$$j(H) = \prod_{\alpha > 0} 2i \operatorname{sen}\left(i\alpha(H)/2\right)$$

$h_i$, $i = 1, \ldots, n$ are the coordinates of $H$ with respect to some orthonormal base of $\mathbf{T}$, and

$$\rho = \frac{1}{2} \sum_{\alpha > 0} \alpha$$

We have,

**Lemma 2.2.** *The function $\theta_e(x) = \theta_e(\exp H)$ is invariant by the Weyl group and*

$$\Delta_{rad}\theta_e^{-1/2}(\exp H) = <\rho, \rho> \theta_e^{-1/2}(\exp H)$$

**Proof.** See [2].

It is now easy to prove,

**Theorem 2.2.** *The fundamental solution of the equation on $G$ is given by*

$$p(t, x, e) = (2\pi t)^{-n/2} \prod_{\alpha > 0} \frac{i\alpha(H)}{2 \operatorname{sen}(i\alpha(H)/2)}$$

$$e^{-\|H\|^2/2t + <\rho,\rho>\frac{t}{2}} \times \mathbf{E}_x(\chi_{\tau > t})$$

*with $x = \exp Ad(g)H \in G - Cut(e)$, for some $g \in G$, and where $\tau$ is, as before, the hitting time of $Cut(e)$ by $(x_s)_{0 \le s \le t}$.*

**Proof.** See [2].

It remains now to calculate the expectation in the last expression for $p(t, x, e)$, which depends very deeply on the geometry of $Cut(e)$ and also on the properties of $(x_s)$.

Following the construction given in [10] for the heat kernel on $S^3$, we can prove, in the particular case of $G = SU(2)$, that

$$p(t, u, I) = \sum_{j \in \mathbf{Z}} \left(\frac{1}{2\pi t}\right)^{3/2} \frac{(4\sqrt{2}j\pi + |\lambda|)}{2\sqrt{2} \operatorname{sen}[4\sqrt{2}j\pi + |\lambda|)/2\sqrt{2}]}$$

$$\times e^{-(4\sqrt{2}j\pi + |\lambda|)^2/2t} e^{t/16}$$

where $u \in SU(2)$, $u \ne I$, $I$ being the identity in $SU(2)$ and $\lambda \in \mathbf{R}$ is such that $|\lambda| = d(u, I)$ and $|\lambda| < 2\sqrt{2}\pi$.

The above sum can be interpreted as a sum over all geodesics in $SU(2)$, between $u$ and $I$, each term depending on the length $4\sqrt{2}j\pi + |\lambda|$ of the corresponding geodesic. See [2] for more details.

We would like to remark that for a general compact Lie group $G$ the heat kernel is given by the following well-known formula (see [4] and [15])

$$p(t, h, e) = \sum_{A \in T_e} \left(\frac{1}{2\pi t}\right)^{n/2} \prod_{\alpha > 0} \frac{i\alpha(H + A)}{2 \, sen \, [i\alpha(H + A)/2]}$$
$$\times \, e^{-\|H+A\|^2/2t + Rt/12}$$

where $T_e = \exp^{-1} e$ is a lattice on $T$, generated by the roots, and as before $x = \exp Adg(H)$; $R$ is the scalar curvature of $G$ that can be given by $R = \langle \rho, \rho \rangle /6$.

The above expression can also be interpreted as a sum over all geodesics, each one corresponding to the vector $H + A$, for each $A \in T_e$. It can be used also to obtain the expectation in Theorem 2.2.

### 2.2. The case of symmetric spaces dual of compact Lie groups

Let $G$ be a compact Lie group and $\mathbf{G}$ its Lie algebra.

The complex Lie group associated to $G$, $G_c$, is a noncompact semi-simple Lie group having $G$ as a maximal compact subgroup. The space $G_c/K$ (where we have put $K = G$) is then a Riemannian symmetric space of noncompact type with Riemannian structure induced by the one of $G$. The exponential map

$$\text{Exp} : T_{\bar{e}}(G_c/K) \to G_c/K$$

is, in this case, a diffeomorphism ($\bar{e} = eK$) and by Theorem 1.1 we can write

$$p(t, \bar{x}, \bar{e}) = (2\pi t)^{-n/2} \prod_{\lambda > 0} \frac{\lambda(\overline{H})}{2 \, sh(\lambda(\overline{H})/2)}$$
$$e^{-d^2(\text{Exp} \, \overline{H}, \bar{e})/2t - \langle \rho, \rho \rangle /2t}$$

where $\bar{x} = \text{Exp}\,(Ad(k)\overline{H})$ for some $k \in K$ and $\overline{H}$ belonging to a maximal abelian subspace of $T_{\bar{e}}(G_c/K)$; $d$ is the Riemannian metric and $\lambda > 0$ are the positive roots of $G_c/K$ with respect to the above maximal abelian subspace.

This expression is very similar to the corresponding one for compact Lie groups reflecting the duality between the two spaces. The existence of only one term in the above formula is a consequence of the fact that there exists only one geodesic between two points in $G_c/K$ (see also [11] and [2]).

## 3. The heat kernel on nilpotent Lie groups

Let us consider now a nilpotent Lie group $G$ with a left invariant Riemannian metric.

As before, we would like to find an explicit expression for the fundamental solution of

$$\left(\frac{\partial}{\partial t} - \frac{1}{2}\Delta\right) u(t, x) = 0$$

for $\Delta$ the Laplace–Beltrami operator associated to that metric.

We observe that for a general left invariant Riemannian metric we have $\exp_e \neq \exp$. So the application of the Elworthy–Truman formula (at least as stated before) is not possible (see [2] for details).

Our method consists of taking a faithful representation of $G$ in some finite dimensional vector space. As $G$ is nilpotent this representation is then given by upper triangular matrices with ones on the diagonal. We denote by $G_N$ the group of such matrices of dimension $N \times N$. In the Lie algebra $\mathbf{G}_N$ of this group we consider the canonical scalar product of matrices

$$< A, B >= N^{-1}\, tr\, AB^t$$

where $tr$ means the trace. We define also in the canonical way the left invariant Riemannian metric associated with this scalar product.

A general element of $G_N$, $x \in G_N$, is a matrix given by $x = (x_{ij})\, i, j = 1, \ldots, N, x_{ij} = 0$ for $i > j$ and $x_{ii} = 1$. Considering the entries $x_{ij}$ as the coordinates on the group we have that the Riemannian metric in these coordinates is given by

$$\gamma_{im,jp}(x) =< \frac{\partial}{\partial x_{im}}, \frac{\partial}{\partial x_{jp}} >= \sum_k (x^{-1})kj(x^{-1})ki\, \delta_{mp}$$

In these coordinates the Laplace–Beltrami operator can be written as

$$\Delta = \sum_{\substack{i,j,m,p \\ i<mj<p}} \gamma^{im,jp}(x)\frac{\partial^2}{\partial x_{im}\partial x_{jp}}$$

It is also possible to write this operator as a sum off squares of vector fields on $G_N$ (see, e.g., [1], [2]).

The Brownian motion on $G_N$, $Z(t) = (Z_{ij}(t))$, $Z(0) = I$, can be defined by the stochastic equation

$$dZ_{ij} = \sum_{i<l<j} Z_{il}\, dB_{1j}$$

where $B_{1j}$ are independent one-dimensional Brownian motions.

As $\Delta$ has no first order term, it can be seen that the Itô equation for $Z(t)$ is the same as the Stratonovich one.

We can now calculate the heat semigroup and its kernel. We have,

**Theorem 3.1.** *Let* $(P_t)_{t\geq 0}$ *the heat semigroup of generator* $(1/2)\Delta$ *on* $G_N$. *Consider a function of the form*

$$f(x) = \prod_{\substack{i,j=1 \\ i<j}}^{N} f_{ij}(x_{ij})$$

*with* $f_{ij}$ *being a Fourier transform of bounded complex measures on* $\mathbf{R}$ *that is*

$$f_{ij}(\lambda) = \int_{\mathbf{R}} e^{i\lambda\alpha} d\mu_{ij}(\alpha)$$

*We have then*

$$(P_t f)(x) = \int_{\mathbf{R}^{N-1}} \prod_{j=1}^{N-1} d\mu_{jN}(\alpha_{jN}) \exp(i \sum_{j=1}^{N-1} \alpha_{jN} x_{jN}).$$

$$\cdot \mathbf{E}_{x^*} \left[ \prod_{1\leq i<j\leq N-1} f_{ij}(Z_{ij}(t)) \exp\left( -\frac{1}{2} \sum_{j,j'=1}^{N-1} \alpha_{jN}\alpha_{j'N} Y_{jj'}(t) \right) \right]$$

*where*

$$Y_{jj'}(t) = \int_0^t \sum_{\substack{1 \\ j\vee j'\leq 1<N}} Z_{jl}(s) Z_{j'l}(s) ds$$

*the expectation being taken with respect to* $(Z_{ij}(t))_{1\leq i<j\leq N-1}$ *and* $x^* = (Z_{ij}(0))_{1\leq i<j\leq N-1}$.

**Proof.** See [1], [2].

If we introduce the measure $\nu_{N(t,x,y)}$ associated to the Brownian bridge

$$Z^y(s) = Z(s) + (s/t)(y - Z(t))$$

and denote by $f_N$ and $P^N t$ the function and semigroup (resp.) defined in $G_N$, we obtain,

**Theorem 3.2.** *Under the same conditions as in the above theorem we can prove that*

$$(P_t^N f_N)(x) = (P_t^{N-1} f_{N-1}^*)(x)$$

*with*

$$f_{N-1}^*(x,y^*) = \int dy d\nu_{N-1}^{(t,x^*,y^*)}(Z^*(.)) f_N(y) [\det 2\pi Y_N]^{-1/2}.$$

$$\exp\left( -\frac{1}{2}(\underline{x}-\underline{y}) Y_N^{-1}(t)(\underline{x}-\underline{y}) \right)$$

*where as before $x^*$ and $y^*$ represent the matrices $x$ and $y$ without the last row and the last column. On the other hand $\underline{x}$ and $\underline{y}$ represent the last column of the corresponding matrices; $Y_N$ is the above matrix $Y$ in the group $G_N$.*

**Proof.** See [1], [2].

From this result we have finally

**Theorem 3.3.** *The heat kernel $p^N(t, x, y)$ on $G_N$ is given by*

$$p^N(t, x, y) = p^{N-1}(t, x^*, y^*) \int d\nu_{N-1}^{(t, x^*, y^*)}(z(\cdot))$$

$$[\det 2\pi Y_N(t)]^{-1/2} \cdot \exp\left(-\frac{1}{2}(\underline{x} - \underline{y})Y_N^{-1}(t)(\underline{x} - \underline{y})\right)$$

*where $p^{N-1}(t, x^*, y^*)$ is the heat kernel in $G_{N-1}$.*

**Proof.** See [1], [2].

We would like to point out that this last result can be integrated to give $p^N(t, x, y)$ as a finite product of Gaussian integrals. Moreover the above formula for the heat kernel is also useful in the study of its asymptotic behavior as $t \downarrow 0$ (see [1] and [2] for some details).

*Acknowledgements.* I would like to thank Professor A. Cruzeiro for a kind invitation to a most interesting Workshop. I would like also to thank Professor Albeverio for giving me this subject, for all the helpful discussions and the constant encouragement. I am also grateful to Professor Elworthy for very interesting discussions.

## REFERENCES

[1] Albeverio S., Arede T. and Haba Z., *On left invariant Brownian motions and heat kernels of nilpotent Lie groups*, Preprint nr. 122, Institut für Mathematik, Ruhr Universitat, Bochum, R. F. A., (1988). To be published in Jour. of Math. Phys.

[2] Arede T., *The heat equation on Lie groups and on some symmetric spaces*, Phd Thesis, Lisbon (1989) (in Portuguese).

[3] Bellaiche C., *Le comportement asymptotique de $p(t, x, y)$ quand $t \to 0$ (points eloignés)*, in Astérisque 84–85, (1981), 151–184.

[4] Benabdallah A., *Noyau de diffusion sur les espaces homogènes compacts*, Bull. Soc. Math. de France, **101**(1973), 263–265.

[5] Berezin F. A., *Laplace operators on semisimple Lie groups*, Amer. Math. Soc. Transl. **21**(2),(1962), 239–339.

[6] Berger M., Gauduchon P. and Mazet E., *Le spectre d'une variété riemannienne*, Lecture Notes in Mathematics, **194**(1971), Springer-Verlag.

[7] Crittenden R., *Minimum and conjugate points in symmetric spaces*, Can. J. Math., **14**(1962), 320–328.

[8] Elworthy D. and Truman A., *The diffusion equation and classical mechanics: an elementary formula*. In: "Stochastic Processes in Quantum Physics," ed. S. Albeverio et. al., Springer-Verlag, Lecture Notes in Physics, **173** (1982), 136–146.

[9] Elworthy K. D., *Stochastic differential equations on manifolds*, London Math. Soc. Lect. Note Series, **70**, Cambridge University Press, (1982).

[10] Elworthy K. D., *The method of images for the heat kernel of $S^3$*, to appear in "Stochastic Processes Geometry and Physics," Proceedings of Ascona Conference, ed. Albeverio et al., World Scientific, (1990).

[11] Eskin L. D., *The heat equation and the Weiertrass transform on certain symmetric Riemannian spaces*, Amer. Math. Soc. Transl., t. 75, (1968), 239–254.

[12] Helgason S., *Differential geometry, Lie groups and Symmetric spaces*, Academic Press, (1978).

[13] Molchanov S., *Diffusion processes and Riemannian geometry*, Russ. Math. Surveys, **30**(1975), 1-36.

[14] Ndumu M., *An elementary formula for the Dirichlet heat kernel on Riemannian manifolds*. In: "From local times to global geometry, control and physics," ed. K. D. Elworthy, Pitman Research Notes in Mathematical series, **150**(1986), Longman, Scientific and Technical.

[15] Urakawa H., *The heat equation on a compact Lie group*, Osaka J. Math., **12**(1975), 285–297.

Maria Teresa Arede
Faculdade de Engenharia - DEMEC
R. dos Bragas, P-4000 Porto
Portugal

# Weak SOBOLEV inequalities.

Dominique Bakry

Laboratoire de Statistique et Probabilités, Université PAUL SABATIER,
118, route de Narbonne, 31062, TOULOUSE Cedex.

## ABSTRACT

A weak SOBOLEV inequality (WSI) is a weakened form of a SOBOLEV inequality associated to a DIRICHLET form. It turns out that it is in fact equivalent to a SOBOLEV inequality. If there is a spectral gap and a (WSI) holds, then we can get a tight weak SOBOLEV inequality (TWSI). Starting with a TWSI, we get upper and lower bounds on the density of the heat semigroup associated with the DIRICHLET form, when $t \to 0$ as well as when $t \to \infty$. In the last chapter, we give a $\Gamma_2$ criterium to get a TWSI on a manifold.

## 1— Weak SOBOLEV Inequalities.

Let us consider a symmetric MARKOV diffusion semigroup $P_t$ on some manifold $E$. Let $\mu$ be it's invariant measure: in the following, we will assume that $\mu$ is a finite measure and, since it is defined up to a constant factor, we may as well assume that $\mu(E) = 1$. We shall write $\langle f \rangle$ for the integral $\int f \, d\mu$ and $\langle f, g \rangle$ for the scalar product $\langle fg \rangle$ of two functions $f$ and $g$ in $\mathbf{L}^2(\mu)$.

The DIRICHLET form associated with $P_t$ is defined by

$$\mathcal{E}(f,f) = \lim_{t \to 0} \frac{1}{2t} \langle f, f - P_t f \rangle,$$

and it's domain $D(\mathcal{E})$ is the space of functions $f$ in $\mathbf{L}^2(\mu)$ for which such a limit exists.

We shall also denote by $E(f)$ the relative entropy of $f^2$, i.e.

$$E(f) = \int f^2 \log f^2 \, d\mu - \langle f^2 \rangle \log \langle f^2 \rangle.$$

Following L.GROSS. we shall say that the semigroup $P_t$ satisfies a logarithmic SOBOLEV inequality if there is a constant $c > 0$ such that

$$\forall f \in D(\mathcal{E}), \ \langle f^2 \rangle = 1 \Rightarrow E(f) \leq c\mathcal{E}(f,f). \qquad (SLog)$$

Here, we are concerned with a another kind of inequality between entropy and energy, namely that there exists three constants $c_1 > 0$, $c_2 > 0$ and $n > 1$ such that

$$\forall f \in D(\mathcal{E}), \ \langle f^2 \rangle = 1 \Rightarrow E(f) \leq \frac{n}{2} \log(c_1 + c_2 \mathcal{E}(f,f)). \qquad (WS)$$

We shall call such an inequality a weak SOBOLEV inequality of constants $c_1$ and $c_2$ and of dimension $n$. The inequality (WS) looks like the inequality (SLog) but is much stronger. In fact, for $n > 2$, it is closely related to the classical SOBOLEV inequality

$$\forall f \in D(\mathcal{E}), \ \langle f^{2n/(n-2)} \rangle^{(n-2)/n} \leq c_1 \langle f^2 \rangle + c_2 \mathcal{E}(f,f). \qquad (S)$$

We recopy from [CS] the following result:

**Lemma 1.1.**—*Inequality (S) implies (WS) with the same constants $n$, $c_1$, $c_2$.*

**Proof.** Let $f$ be a function such that $\langle f^2 \rangle = 1$ and $\nu$ be the probability measure $f^2 . \mu$. We write $\langle f \rangle_\nu$ for the integral of the function $f$ with respect to $\nu$. Set $\varepsilon$ to be $4/(n-2)$, such that $2 + \varepsilon = 2n/(n-2)$. We have

$$E(f) = \langle \log(f^2) \rangle_\nu = \frac{2}{\varepsilon} \langle \log |f|^\varepsilon \rangle_\nu \leq \frac{2}{\varepsilon} \log \langle |f|^\varepsilon \rangle_\nu$$

$$= \frac{2}{\varepsilon} \log(\langle |f|^{2+\varepsilon} \rangle) = \frac{n}{2} \log(\langle |f|^{\frac{2n}{n-2}} \rangle^{\frac{n-2}{n}}).$$

Then, we write use the (S) inequality to write

$$\log(\langle |f|^{\frac{2n}{n-2}} \rangle^{\frac{n-2}{n}}) \leq \log(c_1 + c_2 \mathcal{E}(f,f)),$$

and we get the result.                                                              □

It is more surprising (but still true) that the inequality (WS) implies the inequality (S) with the same constant $n$, but with constants $c_1$ and $c_2$ which may be different. (We will see that later.)

Assume that (WS) holds. Then. applying it to the constant function $f = 1$, we see that $c_1 \geq 1$. We shall say that (WS) is tight if it holds with $c_1 = 1$. (In what follows, we shall refer to a tight weak SOBOLEV inequality as (TWS).)

**Lemma 1.2.—** *If (TWS) holds. then it implies (SLog) with $c = \frac{n}{2}c_2$.*

**Proof.** Just use the inequality $\log(1 + x) \leq x$. ◻

In fact, as soon as (WS) holds with some constants $c_1 > 1$ and $c_2$, it holds with $c_1 = 1$ if we assume moreover that the DIRICHLET form $\mathcal{E}$ has a spectral gap $\lambda$ :

$$\forall f \in \mathbf{L}^2(\mu), \ \langle f^2 \rangle \leq \langle f \rangle^2 + \frac{1}{\lambda}\mathcal{E}(f, f). \qquad (G)$$

($\lambda$ is the first non 0 eigenvalue of the generator of $P_t$.)

We will proove in fact a more general result a more general result : suppose we have a general inequality

$$\forall f \in \mathbf{L}^2(\mu), \ \langle f^2 \rangle = 1 \Rightarrow \leq E(f) \leq \Phi(\mathcal{E}(f, f)), \qquad (S\Phi)$$

with some concave increasing function $\Phi$ on $[0, \infty)$. Then, we have

**Lemma 1.3.—** *If (G) and $(S\Phi)$ hold. then $(S\hat{\Phi})$ holds, with $\hat{\Phi}(x) = \inf(\Phi(x), Kx)$, where $K = (2 + \Phi(\lambda))/\lambda$.*

**Proof.** We will use the following inequality of DEUSCHEL and STROOCK ([DSt], p. 246) : if $\hat{f}$ denotes the function $f - \langle f \rangle$, then

$$E(f) \leq E(\hat{f}) + 2\langle \hat{f}^2 \rangle.$$

Assume then that $(S\Phi)$ holds. We may write

$$H(f) \leq H(\hat{f}) + 2\langle \hat{f}^2 \rangle \leq \langle \hat{f}^2 \rangle [2 + \Phi(\frac{\mathcal{E}(f, f)}{\langle \hat{f}^2 \rangle})]$$

$$\leq \mathcal{E}(f, f)\Psi(\frac{\mathcal{E}(f, f)}{\langle \hat{f}^2 \rangle}),$$

with $\Psi(x) = (2 + \Phi(x))/x$. Then. we use the fact that $\frac{\mathcal{E}(f,f)}{\langle \hat{f}^2 \rangle} \geq \lambda$, where $\lambda$ is the spectral gap and we get

$$H(f) \leq K\mathcal{E}(f, f),$$

where $K = \sup_{x \geq \lambda} \Psi(x) = \Psi(\lambda)$. (Here, the last equality comes from the convexity of $\Phi$.) Hence, we may replace $\Phi$ by $\hat{\Phi}$ in $(S\Phi)$. ◻

Now, when $\Phi(x) = \frac{n}{2}\log(c_1 + c_2 x)$, for every $K$ we can find a constant $c_2'$ such that $\frac{n}{2}\log(1 + c_2' x) \geq \inf\{\Phi(x), Kx\}$. This gives the result.

On another hand. when the function $\Phi$ in $(S\Phi)$ is such that $\Phi(0) = 0$ and $\Phi(x) \leq Kx$, for some $K > 0$. then the argument of lemma (1.2) applies and shows that (SLog) holds with constant $c = K$.

But, as it is well known. if we replace in inequality (SLog) the function $f$ by $(1 + tf)/\langle(1 + tf)^2\rangle^{1/2}$, and if we let $t$ tend to 0, we get in the limit the inequality (G) with spectral gap $\lambda = 2/c$.

To summarize. having $(S\Phi) + (G)$ is equivalent to have $(S\Phi)$ with $\Phi(0) = 0$ and $\Phi(x) \le Kx$.

In what follows. we are mainly interested in the case where $E$ is a smooth compact manifold, and where $P_t$ is the heat semigroup associated to an elliptic second order differential operator $L$. without constant term : in local coordinates

$$L = \sum_{ij} g^{ij}(x)\frac{\partial^2}{\partial x^i \partial x^j} + \sum_i b^i(x)\frac{\partial}{\partial x^i},$$

and $P_t = \exp(tL)$. Since we restrict ourselves to the case where $P_t$ is symmetric, we may rewrite $L$ under the form

$$L = \Delta + \nabla h.$$

where $\Delta$ is the LAPLACE-BELTRAMI operator associated with the Riemannian structure $g^{ij}$, and $e^h$ is the density of the measure $\mu$ with respect to the Riemannian measure. In this case, the DIRICHLET form is

$$\mathcal{E}(f, f) = \int |\nabla f|^2 \, d\mu.$$

where the term $|\nabla f|$ is the lenght of the gradient of $f$ in the Riemannian metric. In this context, we know that there always exists a SOBOLEV inequality with dimension $n$ being the dimension of the space $E$, and also a spectral gap $\lambda$. So we know a (TWS) inequality holds with dimension $n$. In fact, from what follows, we will see that one cannot expect a dimension in (WS) lower than the dimension of $E$. but it could be useful to consider (WS) inequalities for larger dimensions : for example, when $L$ is the LAPLACE-BELTRAMI operator of the $n$-dimensionnal sphere. then we will show that the best constant in the logarithmic SOBOLEV inequality may be deduced from a (TWS) one of dimension larger than the dimension of the sphere.

## 2— Bounds on the heat-semigroup.

Many authors have studied the relationship between inequalities like (SLog) or (S) and bounds on the semigroup $P_t$ associated with the DIRICHLET form $\mathcal{E}(f, f)$. (See for example [G], [CSK], [DS], [V], and the books of E.B.DAVIES [D] and DEUSCHEL-STROOCK [DSt].)

The key theorem is GROSS' one [G]: if $(SLog)$ holds, then the operator $(P_t)$ maps $\mathbf{L}^p(\mu)$ into $\mathbf{L}^q(\mu)$ with norm 1 as soon as $p \geq 1$ and $q \geq 1 + (p - 1) \exp(\frac{4t}{c})$.

When a SOBOLEV inequality holds, then the above result may be improved: for every $t > 0$, $(P_t)$ maps $\mathbf{L}^1(\mu)$ into $\mathbf{L}^\infty(\mu)$, with a norm bounded for $0 < t \leq 1$ by $Ct^{-n/2}$, with some constant $C = C(c_1, c_2, n)$, $n$ being the dimension in the SOBOLEV inequality. This has been prooved by various methods in [D] (using a version of weak SOBOLEV inequalities), in [CSK] (using a NASH inequality, which is a consequence of a SOBOLEV one), and in [V].

The result of [V] is even stronger: there is equivalence between a SOBOLEV inequality and the above bound on the norm of $(P_t)$ between $\mathbf{L}^1(\mu)$ and $\mathbf{L}^\infty(\mu)$. In fact, the bound $\|P_t\|_{1,\infty} \leq Ct^{-n/2}$ implies a SOBOLEV inequality with dimension $n$, but when $C = C(c_1, c_2, n)$, we dot not get the same constants $c_1$ and $c_2$ in $(S)$: in the equivalence between the above majorization on $P_t$ and a SOBOLEV inequality, we loose nothing on the dimension $n$ but we loose on the constants $c_1$ and $c_2$.

DAVIES' method shows that the same is true for a weak SOBOLEV inequality: combining this with VAROPOULOS' one, we get the equivalence between (WS) and (S), up to the constants $c_1$ and $c_2$.

The above bound on $\|P_t\|_{1,\infty}$ shows that the operator $(P_t)$ has a density $p_t(x, y)$ with respect to the invariant measure $\mu$,

$$P_t(f)(x) = \int_E p_t(x, y) f(y) \mu(dy),$$

with a uniform bound on $p_t(x, y)$ given by

$$|p_t(x, y)| \leq Ct^{-n/2}. \tag{2.1}$$

DAVIES improved this result to show that in fact

$$p_t(x, y) \leq C(n) t^{-n} \exp\{\frac{-d^2(x, y)}{4t}\}, \tag{2.2}$$

where the distance $d(x, y)$ may be defined in terms of the generator $L$ of $(P_t)$ by the following procedure:
define first the squared field operator $\Gamma$ by

$$\Gamma(f, f) = \frac{1}{2}(L(f^2) - 2fL(f)) = \lim_{t \to 0} \frac{1}{2t}\{P_t(f^2) - (P_t(f)^2\}.$$

Because of the symmetry assumption, we have

$$\mathcal{E}(f,f) = \int_E \Gamma(f,f)\, d\mu.$$

Also, $\Gamma$ is positive because of the MARKOV property of the semigroup $(P_t)$. This allows us to introduce the distance

$$d(x,y) = \sup_{\Gamma(f,f)\leq 1} (f(x) - f(y)).$$

This is the distance appearing in (2.2). Of course, if we are dealing with the case of an elliptic operator on $E$, with $L = \Delta + \nabla h$, then $\Gamma(f,f) = |\nabla f|^2$ and $d(x,y)$ is nothing else but the Riemannian metric associated with the second order part of $L$.

Notice the difference in the exponant of $t$ between the off diagonal estimate (2.2) and the uniform one (2.1) (which is in fact an estimate on the diagonal): it is well known that, when $(P_t)$ is the semigroup of the Brownian motion on a compact Riemannian manifold, one has an estimate of the form

$$p_t(x,y) \approx C(x,y)t^{-n/2}\exp\{\frac{-d^2(x,y)}{4t}\}$$

in a neighborhood of the diagonal of $E \times E$: more precisely, this holds when $y$ is not in the cut locus of $x$. But, if we look for example at the Brownian motion on the $n$ dimensionnal sphere, then, for two opposite points $x$ and $y$, we get

$$p_t(x,y) \approx Ct^{-n+\frac{1}{2}}\exp\{\frac{-d^2(x,y)}{4t}\}.$$

In some sense, this seems to be the worst case, and one would expect the exponant $t^{-n+1/2}$ in the majorization (2.2) instead of $t^{-n}$.

Using the distance $d(x,y)$ associated with $L$, we may introduce the diameter of $E$, which is

$$\mathrm{diam}(E) = \sup_{(x,y)} d(x,y).$$

We are concerned here with the case where a (TWS) inequality holds, with dimension $n$ and some constant $c_2$. In this case, we already know that there is a gap in the spectrum of $L$, and so we know that some kind of ergodicity of the semigroup holds. It is not surprising then that we are able to get lower bounds on the density $p_t(x,y)$ of the semigroup.

Let us state our main results:

**Theorem 2.1.**—$\operatorname{diam}(E) \leq \sqrt{nc_2}\frac{\pi}{2}$

This first result shows that it is useless to look for a (TWS) inequality on a non compact manifold if this manifold is complete for the Riemannian metric associated with $L$.

**Theorem 2.2.**—*Uniform upper bound : let $F_1(\tau)$ be the function defined for $\tau > 0$ by*

$$F_1(\tau) = \frac{1}{\sqrt{1+\tau}}\operatorname{argth}(\frac{1}{\sqrt{1+\tau}}).$$

*This function decreases from $\infty$ to $0$ when $\tau$ increases from $0$ to $\infty$. Let $G_1(\tau)$ be it's inverse function : $G_1 \circ F_1(\tau) = \tau$. Then*

$$p_t(x,y) \leq 2^{-n}G_1(\frac{2t}{nc_2})^{n/2}\exp\{\frac{t}{c_2}[2 + G_1(\frac{2t}{nc_2})]\}.$$

**Theorem 2.3.**—*Off diagonal majorization : let $F_2(\tau)$ be the function defined for $\tau > 0$ by*

$$\begin{cases} F_2(\tau) = \dfrac{\operatorname{argth}(\sqrt{1-\tau})}{\sqrt{1-\tau}} & \text{for} \quad 0 < \tau < 1 \\[2mm] F_2(\tau) = \dfrac{\operatorname{arctg}(\sqrt{\tau-1})}{\sqrt{\tau-1}} & \text{for} \quad \tau > 1 \end{cases}$$

*Set $F(\alpha,\tau)$ to be the function*

$$F(\alpha,\tau) = (1 + \alpha^2 + \alpha\sqrt{\tau})F_2\{(1+\alpha^2)(\frac{\alpha + \sqrt{\tau}}{1 + \alpha^2 + \alpha\sqrt{\tau}})^2\}.$$

*Then, for every $\alpha > 0$, this is a decreasing function of $\tau > 0$ : for fixed $\alpha$, let $G_2(\alpha,t)$ be it's inverse function. Then, setting*

$$\alpha = \operatorname{tg}(\frac{\pi}{2} - \frac{d(x,y)}{\sqrt{nc_2}}),$$

*which is always positive because of theorem (2.1), we get*

$$p_t(x,y) \leq \left\{\frac{\alpha + \sqrt{G_2(\alpha,\frac{2t}{nc_2})}}{2}\right\}^n \exp\{\frac{t}{c_2}[2 + \alpha^2 - G_2(\alpha,\frac{2t}{nc_2})]\}.$$

It is not easy to see (but true) that, when $d(x,y)$ goes to $0$, this goes to the uniform upper bound.

**Theorem 2.4.**—*Uniform lower bound : let $F_2(\tau)$ be the function of theorem (2.3) and $G_3$ be it's inverse function. Then*

$$p_t(x,y) \geq 2^{-n} G_3(\frac{2t}{nc_2})^{n/2} \exp\{-\frac{t}{c_2}[G_3(\frac{2t}{nc_2} - 2]\}.$$

**Remark.—**

When $\alpha = 0$, the right hand sides of the two inequalities in theorems (2.4) and (2.3) are equal. This means the following thing: assume that there is equality in theorem (2.1) (diam$(E) = \sqrt{nc_2}\frac{\pi}{2}$). If we choose two opposite points on $E$, i.e. two points $x$ and $y$ for which $d(x,y) = $ diam$(E)$, then the inequalities in theorems (2.3) and (2.4) are equalities : so we get an explicit formula for the lower bound of $p_t(x,y)$, for every $t$. Unfortunately, we will see later that in fact, the inequality in theorem (2.1) is always strict when $P_t$ is the heat semigroup of a Riemannian manifold.

The formulae in theorem (2.2), (2.3), and (2.4) are not easy to handle. So, it is useful to get asymptotic formulae when $t \to 0$ and $t \to \infty$. This can be easily done by taking asymptotic expansions of the functions $G_i$. We get

**Corollary 2.5.**—*Uniform estimates at $t = 0$ : when $t \to 0$,*

$$p_t(x,y) \leq (\frac{e}{4})^{-n/2}(\frac{nc_2}{2t} - \frac{3}{2})^{-n/2} + o(t), \text{ and}$$

$$p_t(x,y) \geq (\frac{n\pi e c_2}{8t})^n \exp(-\frac{\pi^2 n^2 c_2}{16t})(1 + o(t)).$$

**Corollary 2.6.**—*Off diagonal estimate at $t = 0$ : when $t \to 0$,*

$$p_t(x,y) \leq C(d(x,y), n, c_2)t^{-n} \exp(-\frac{d^2(x,y)}{4t})(1 + o(t)).$$

**Remark.—**

If we are in the case where $\text{diam}(E) = \sqrt{nc_2}\frac{\pi}{2}$, then, for $d(x, y) = \text{diam}(E)$, this estimate gives,

$$p_t(x, y) \approx Ct^{-n} \exp(-\frac{d^2(x, y)}{4t}). \tag{2.1}$$

But we know that, on a complete Riemannian manifold, the worst possible case is

$$p_t(x, y) \approx Ct^{-n_0 + \frac{1}{2}} \exp(-\frac{d^2(x, y)}{4t}),$$

where $n_0$ is the dimension of the space. On the other hand, on the diagonal,

$$p_t(x, x) \approx C(x)t^{-n_0/2}.$$

Comparizon of the second formula with the upper bound shows that $n \geq n_0$, and the first formula shows then that (2.1) cannot hold. So, in the case of Riemannian manifolds, the inequality of theorem (2.1) is always strict.

**Corollary 2.7.**—*Uniform estimates when* $t \to \infty$ :

$$(1 + o(1)) \exp\{-\frac{4t}{c_2} \exp(-\frac{4t}{nc_2})\} \leq p_t(x, y) \leq (1 + o(1)) \exp\{+\frac{4t}{c_2} \exp(-\frac{4t}{nc_2})\}.$$

This last corollary shows the rate of convergence of the semigroup to 1, when $t \to \infty$. In practise, this is perhaps the more useful result.

**Sketch of the proofs.**

The proofs of the preceeding theorems rely on an idea of DAVIES. This idea is quite simple, but the computation are rather technical. So, we will just here show the idea, and refer the reader to [BM] for the details.

First of all, notice that if we change the semigroup $P_t$ into the semigroup $P_{c_2 t}$, everything is preserved, except that the new semigroup satisfies a weak SOBOLEV inequality with $c_2 = 1$. So, from now on, we will restrict ourselves to this later case.

We begin with the uniform bounds. Everything goes back to GROSS' method. Recall that $L$ is the generator of $P_t$ in $\mathbf{L}^2(\mu)$. Consider a function $f$ on $E$, in the domain of $L$, and satisfying $0 < a < f < b < \infty$, for two constants $a$ and $b$. The differences of two such functions are dense in $\mathbf{L}^2(\mu)$: take for example $f = P_\varepsilon g$, where $0 < a < g < b < \infty$ and let $\varepsilon$ go to 0.

We consider then the function $\hat{f}(t) = P_t f$ : this function is bounded above and away from 0, and satisfies the heat equation

$$\frac{\partial}{\partial t}\hat{f} = L\hat{f}; \ \hat{f}(0) = f.$$

Consider two smooth functions $q(t)$ and $\hat{m}(t) : \mathcal{R} \to \mathcal{R}$, with $q(0) = 1$, $\hat{m}(0) = 0$ and form the quantity

$$U(t) = \exp(-\hat{m}(t))\langle \hat{f}(t)^{q(t)}\rangle^{1/q(t)}.$$

Then, if we take the derivative of $U$, using the heat equation, we get

$$\frac{dU}{dt} = \frac{U(t)}{\langle \hat{f}^{q(t)}\rangle} \frac{q'(t)}{q^2(t)} K(\hat{f}, q, m),$$

where

$$K(f, q, m) = \langle f^q \log f^q\rangle - \langle f^q\rangle \log\langle f^q\rangle - \frac{q^2}{q'}\{-\langle f^{q-1}, Lf\rangle + \hat{m}'\langle f^q\rangle\}.$$

Now, assume that we can find two functions $c(q)$ and $m(q)$ such that, for every function $f$ in the domain of $L$, bounded above and away from 0, the following holds

$$\langle f^q \log f^q\rangle - \langle f^q\rangle \log\langle f^q\rangle - c(q)\{-\langle f^{q-1}, Lf\rangle + m(q)\langle f^q\rangle\} \le 0. \qquad (Slog(p))$$

Then, if we choose $\hat{m}(t)$ and $q(t)$ such that

$$\frac{d\hat{m}}{dt} = m(q(t)) \text{ and } \frac{q^2}{q'(t)} = c(q(t)), \qquad (E)$$

we get, if $q' \ge 0$,

$$\langle \hat{f}(t)^{q(t)}\rangle^{1/q(t)} \le \exp(\hat{m}(t))\langle f\rangle.$$

Moreover, if we can manage in the above expression such that $q(t_0) = \infty$, we get

$$\|P_{t_0} f\|_\infty \le \exp(\hat{m}(t_0))\|f\|_1,$$

which shows that the operator $P_t$ being bounded from $\mathbf{L}^1(\mu)$ into $\mathbf{L}^\infty(\mu)$ has a density $p_t(x, y)$ bounded above by $\exp(m(t_0))$.

On the other hand, if we have $q' \le 0$, then we get

$$\langle \hat{f}(t)^{q(t)}\rangle^{1/q(t)} \ge \exp(\hat{m}(t))\langle f\rangle.$$

If we have $q(t_0) = -\infty$, then we get for the same reason

$$p_t(x, y) \ge \exp(\hat{m}(t_0)).$$

The system (E) can be written

$$dt = \frac{c(q)}{q^2}\, dq, \quad d\hat{m} = \frac{m(q)c(q)}{q^2}\, dq,$$

so that, if $c(q) > 0$ on $q > 1$, and

$$t_0 = \int_1^\infty \frac{c(q)}{q^2}\, dq,$$

we have an upper bound $\hat{M}_0$ of $\log p_{t_0}(x, y)$ equal to

$$\hat{M}_0 = \int_1^\infty \frac{m(q)c(q)}{q^2} \, dq.$$

Also, if $c(q) < 0$ on $q < 1$ and

$$t_0 = -\int_{-\infty}^1 \frac{c(q)}{q^2} \, dq,$$

we get a lower bound $\hat{m}_0$ of $\log p_{t_0}(x, y)$ equal to

$$\hat{m}_0 = -\int_{-\infty}^1 \frac{m(q)c(q)}{q^2} \, dq.$$

The problem is now to check inequality (SLog(p)). Starting with the fact that the function $\log(1 + x)$ is convex, we see that in fact (WS) is equivalent to the following family of inequalities

$$\forall x > 0, \ \forall f \in D(\mathcal{E}), \ \langle f^2 \rangle = 1 \Rightarrow E(f) \le \frac{n}{2} \{\log(1 + x) + \frac{1}{1 + x}(\mathcal{E}(f, f) - x)\}.$$

(We just replaced the concave function $\log(1 + x)$ by the family of it's tangent lines.) For a general $f$, this becomes

$$\forall x > 0, \ \forall f \in D(\mathcal{E}), \ E(f) \le \frac{n}{2} \{(\log(1 + x) - \frac{x}{1 + x}) \langle f^2 \rangle + \frac{1}{1 + x} \mathcal{E}(f, f)\}.$$

Now, apply this inequality to the function $g = f^{q/2}$: we have

$$E(g) = \langle f^q \log f^q \rangle - \langle f^q \rangle \log \langle f^q \rangle.$$

On the other hand, we also have

$$\mathcal{E}(f^{q/2}, f^{q/2}) = -\frac{q^2}{4(q - 1)} \langle f^{q-1}, Lf \rangle.$$

(Here, we use the fact that we are dealing with a diffusion semigroup so that a change of variables formula applies on $L$.) In the end, we get a whole family of inequality (SLog(q)), with a parameter $x > 0$ and coefficient

$$\begin{cases} c(q) & = c(q, x) & = & \dfrac{q^2}{4(q - 1)} \dfrac{n}{2(1 + x)} \\ m(q) & = m(q, x) & = & \dfrac{4(q - 1)}{q^2} \{(1 + x) \log(1 + x) - x\}. \end{cases}$$

Remark that the function $(q-1)c(q,x)$ is positive, and so we may apply our method. It just remains to find a function $x(q) > 0$ such that the system

$$\begin{cases} t_0 & = \int_1^\infty C(q,x(q))\,\dfrac{dq}{q^2} \\[2mm] M_0 & = \int_1^\infty C(q,x(q))m(q,x(q))\,\dfrac{dq}{q^2} \end{cases}$$

gives an optimal result. The miracle is that the computation can be made explicitly and gives the result of theorem (2.2). The same is true for the lower bound, we just have to exchange the intervals $[1,\infty)$ with $(-\infty,1]$, and we get theorem (2.4).

To deal with the off diagonal estimate, we use again a idea of DAVIES: we take a smooth function $\varphi$ and consider the semigroup $P_t^\varphi$ given by $P_t^\varphi(f) = \exp(-\varphi)P_t(\exp(\varphi)f)$. It is cleat that, if $p_t(x,y)$ is the density of $P_t$, then $P_t^\varphi$ has density $\exp(\varphi(x)-\varphi(y))p_t(x,y)$. This semigroup is not MARKOV in general, but we may apply as well the GROSS' method. Let $L^\varphi$ be it's generator: using the majorization

$$\Gamma(f,\varphi)^2 \le \Gamma(f,f)\Gamma(\varphi,\varphi)$$

and a few manipulations, it is not hard to show that the following inequalities hold, for every $b > 1$

$$\begin{cases} -\langle f^{q-1}, Lf\rangle & \le -b\langle f^{q-1}, L^\varphi f\rangle + \rho(b,q)\langle f^q, \Gamma(\varphi,\varphi)\rangle & \text{if } q > 1, \\[2mm] -\langle f^{q-1}, Lf\rangle & \ge -b\langle f^{q-1}, L^\varphi f\rangle + \rho(b,q)\langle f^q, \Gamma(\varphi,\varphi)\rangle & \text{if } q < 1, \end{cases}$$

where $\rho(b,q) = b(1 + \dfrac{b(q-2)^2}{4(q-1)(b-1)})$. This gives rise to inequalities (SLog(q)) for the operator $L^\varphi$, using only the constant $h = \|\Gamma(\varphi,\varphi)\|_\infty$. These inequalities involve now two parameters $b > 1$ and $x > 0$ instead of $x$ alone. We may apply the same method in order to get the majorization

$$\log(p_{t_0}(x,y)) + \varphi(x) - \varphi(y) \le M(t_0,h),$$

with a function $M(t_0,h)$ given by a variationnal problem which can be solved explicitly. The last inequality may be read as

$$\log(p_{t_0}(x,y)) \le M(t_0,h) + \varphi(y) - \varphi(x).$$

Now, if we take the infimum of the right hand side over all the functions $\varphi$ such that $\Gamma(\varphi,\varphi) \le h$, and if we remember that $d(x,y) = \sup_{\Gamma(\varphi,\varphi)\le 1}(\varphi(x) - \varphi(y))$, we get

$$\log(p_{t_0}(x,y)) \le M(t_0,h) - \sqrt{h}d(x,y).$$

It remains to take the infimum of the right hand side over all the $h > 0$ to get the off diagonal estimate of theorem (2.3). (Once again, the computation can be made explicitly until the end, and (2.3) gives the result of the computation.)

We may want to play the same game for the lower bounds: we get

$$\log(p_{t_0}(x,y)) \geq m(t_0,h) + \sqrt{h}d(x,y).$$

Unfortunately, the function $m(t_0,h)$ that we obtain has the form $-\sqrt{h}(n\pi/2) + m(t_0,0)$. If we take the maximum over all the values of $h$, we get that this is infinite if $d(x,y) > n\pi/2$: from this follows theorem (2.1) on the diameter. Then, we just get the same inequality than the one for $h = 0$, so that the minorization in this case is just the uniform minorization. It would be interesting to know if there is a similar method which gives non uniform lower bounds. ⬜

## 3— A $\Gamma_2$ criterium for weak SOBOLEV inequalities.

From now on, our space will be a compact $C^\infty$ $p$-dimensionnal manifold $E$, and the operator $L$ will be elliptic, with smooth coefficients: $L = \Delta + \nabla h$, where $\Delta$ is the LAPLACE-BELTRAMI operator associated with some Riemannian structure $g$ on $E$. If $\nu(dx)$ is the RIEMANN measure, then the measure $\mu(dx)$ will be $c \exp h(x)\mu(dx)$, where $c$ is a normalizing constant such that $\mu(E) = 1$. Since $E$ is compact, we already know that some SOBOLEV inequality holds with dimension $n = p$. We also know that $L$ has a discrete spectrum. Hence the spectral gap inequality (G) also holds, so that there is no question about the existence of a (TWS) inequality with dimension $p$. Moreover, we know from the previous chapter that, if a (TWS) inequality holds with dimension $n$ and constant $c_2$, then $n \geq p$ and $c_2 \geq \dfrac{4\text{diam}(E)^2}{n\pi^2}$, where $\text{diam}(E)$ is the Riemannian diameter. At least in the case where $h = 0$, we also know that the previous inequality is strict.

Let's look first at the case where $E$ is the $p$-dimensionnal sphere $S_p$, i.e. the sphere of radius 1 in $\mathcal{R}^{p+1}$. $L$ is the usual LAPLACE-BELTRAMI operator of $S_p$, and the measure $\mu$ is the surface measure normalized to a probability measure. Then, the previous inequality reads $c_2 \geq c_2(opt) = 4/n^2$.

On the other hand, T.AUBIN has computed in this case the best constants $c_1$ and $c_2$ appearing in the SOBOLEV inequality (S) for $n = p$: he found ([A], p.50)

$$c_1 = 1, \quad c_2 = c_2(S) = \frac{4}{p(p-2)}.$$

In this chapter, our purpose, among other things, is to show that we can get a better (WS) inequality than the one which can be deduced from AUBIN's one: namely, a (TWS) inequality with dimension $n = p$ and with $c_2 = c_2(WS) = \dfrac{4}{p(p-1)}$. We do not know if this constant is the best possible for (TWS) inequalities on the sphere.

Our method is completely different from AUBIN's one, and very close to the method used in [BE1], [BE2] and [BE3] to get usual and logarithmic SOBOLEV inequalities. It generalizes to the case of a general $L = \Delta + \nabla h$ under the same hypothesis on the curvature that we shall just describe. Recall that $g$ is the RIEMANN metric associated with $L$; let $\nabla$ be the Riemannian connection, and Ric be the RICCI tensor of the Riemannian structure. Then we call Ric($L$) the symmetric tensor Ric $-\nabla\nabla h$. Our main hypothesis (which we call (CD)-hypothesis) is the following:

there exists 2 constants $n \geq p$ and $\rho > 0$ such that

$$0 \leq -\nabla h \otimes \nabla h + (n - p)[\text{Ric}(L) - \rho g], \qquad (CD)$$

this inequality holding in the sense of symmetric tensors, which means that in a local system of coordinates, this tensor is given by a positive symmetric matrix, at every point $x$ of $E$.

Notice that this inequality holds for $n = p$ only in the case where $L = \Delta$. Also, when $L = \Delta$, this inequality holds for a given $\rho$ iff $\rho \geq \inf_E \text{Ric}$: from there comes the name (CD), which stands for Curvature-Dimension.

Another example one can look at is when $L$ is a STURM-LIOUVILLE operator on a real interval (which is not a compact manifold but this is irrelevant for the moment): if $Lf(x) = f''(x) - a(x)f'(x)$, then (CD) is equivalent to

$$(n - 1)(a' - \rho) \geq a^2.$$

**Theorem 3.1.**—*If (CD) holds, then a (TWS) inequality holds with dimension $n$ and $c_2 = \frac{4}{n\rho}$.*

For example, on the $p$-dimensionnal sphere, (CD) holds with $n = p$ and $\rho = n - 1$, and we get the announced result.

In fact, under the same hypothesis, we get a whole family of of (TWS) inequalities depending on a parameter $\alpha$.

Let us describe the results: we set

$$\begin{aligned}
b_0(\alpha) &= (n - 1)(n - 16)\alpha^2 - 2\alpha(n - 7) + n(n - 1) ; & (3.1) \\
b_1(\alpha) &= 3\alpha(n - 4) - 4(n - 1); & (3.2) \\
b_2(\alpha) &= \alpha(n - 7) - (n - 1); & (3.3) \\
b_3(\alpha) &= n - \alpha(n - 1), \text{ and} & (3.4)
\end{aligned}$$

$$Q(\alpha, X) = (b_0^2 - 4Xb_1)^2 - 64Xb_3b_2^2. \qquad (3.5)$$

For every $\alpha$ in $[1, n/(n - 1)]$, the second degree equation $Q(\alpha, X) = 0$ has 2 real roots between 0 and 1. Then, we call $X(\alpha)$ the largest one. (The point $(\alpha, X(\alpha))$ in the plane is on an algebraic curve of degree 4.) We have

**Theorem 3.2.**—*If (CD) holds, then, for every $\alpha$ in $[1, n/(n-1)]$, the operator $L$ satisfies a (TWS) inequality with dimension $n/X(\alpha)$ and $c_2 = \dfrac{4X(\alpha)}{n\alpha\rho}$.*

The function $\alpha \to X(\alpha)$ decreases from $X(1) = 1$ to $X(\dfrac{n}{n-1}) = \dfrac{n(4n-1)}{4(n+2)^2}$. The slope of the curve is $0$ for $\alpha = 1$ and $-\infty$ for $\alpha = n/(n-1)$.

For $\alpha=1$, we recover the previous theorem. For other values of $\alpha$, we loose on the dimension in the (TWS) inequality, but we gain on the value of $c_2$.

If we use the inequality $\log(1+x) \leq x$, for every value of $\alpha$, we get a (SLog) inequality with constant $c = 2/(\alpha\rho)$. This inequality is optimum when $\alpha = n/(n-1)$. For the sphere, we recover in this way the best constant in logarithmic SOBOLEV inequalities: so, it appears that this (SLog) inequality is in fact a consequence of a (TWS) inequality with dimension $4(n+2)^2/(4n-1)$.

## Sketch of the proof.

Once again, we will just describe the idea, which is quite simple, but leave the computations to the reader (or refer to [B2]). First, let us describe where the (CD) hypothesis comes from: we already introduced the operator $\Gamma$:

$$2\Gamma(f,f) = L(f^2) - 2fLf.$$

By construction, it satisfies $\langle \Gamma(f,f) \rangle = -\langle f, Lf \rangle = \mathcal{E}(f,f)$, and if we compute it's value, we get $\Gamma(f,f) = |\nabla f|^2$. We will repeat this construction procedure, but replacing the bilinear form $f \to f^2$ by $f \to \Gamma(f,f)$. Then we get

$$2\Gamma_2(f,f) = (\mathrm{def}) = L\Gamma(f,f) - 2\Gamma(2, Lf).$$

Once again, by construction, we have $\langle \Gamma_2(f,f) \rangle = \langle Lf, Lf \rangle$. On the other hand, if we compute it's value on a given function $f$, we get

$$\Gamma_2(f,f) = |\nabla\nabla f|^2 + \mathrm{Ric}(L)(df, df),$$

where $\mathrm{Ric}(L)$ is the tensor introduced in the definition of (CD), and $|\nabla\nabla f|$ denotes the HILBERT-SCHMIDT norm (in the Riemannian metric) of the symmetric tensor $\nabla\nabla f$. Now, using the fact that $\Delta f$ is the trace of $\nabla\nabla f$, a simple computation shows that the (CD) hypothesis is in fact equivalent to the following inequality, valid for every smooth function on $E$,

$$\Gamma_2(f,f) \geq \frac{1}{n}(Lf)^2 + \rho\Gamma(f,f).$$

This notion of RICCI curvature and dimension of a diffusion semigroup has been developed in [B1] and in [BE4], and we refer the interested reader to these papers.

In order to understand the role of the $\Gamma_2$ operator in the computation on the semigroups, we will start with the same kind of computation than in GROSS' method.

We consider a smooth function $f$ on $E$, bounded above and below by positive constants, and such that $\langle f \rangle = 1$. Then, we set $\hat{f}(t) = P_t f$. Since the measure $\mu$ is invariant, we have $\forall t$, $\langle \hat{f}(t) \rangle = 1$. Also, since the only invariant functions on $E$ are the constants, we will have $\hat{f}(t) \to \langle f \rangle = 1$ in $\mathbf{L}^2(\mu)$ when $t \to \infty$. Take a smooth function $\Phi : \mathcal{R}_+ \to \mathcal{R}_+$ and set $U(t) = \langle \Phi(\hat{f}(t)) \rangle$. Then, using the diffusion property of $L$ and the symmetry assumption, we have

$$
\begin{cases}
U'(t) &= -\langle \Phi''(\hat{f}(t)), \Gamma(\hat{f}(t), \hat{f}(t)) \rangle; \\[2mm]
U''(t) &= 2\langle\langle \Phi''(\hat{f}(t)), \Gamma_{\!2}(\hat{f}(t), \hat{f}(t)) \rangle + 2\langle \Phi^{(3)}(\hat{f}(t)), \Gamma(\hat{f}(t), \Gamma(\hat{f}(t), \hat{f}(t))) \rangle + \\
& \quad \langle \Phi^{(4)}(\hat{f}(t)), \Gamma^2(\hat{f}(t), \hat{f}(t)) \rangle.
\end{cases}
$$

From now on, we will take $\Phi(x) = x \log x$, so that we have, with $g$ standing for $\hat{f}(t)$,

$$
\begin{cases}
U'(t) &= \langle g^{-1}, \Gamma(g,g) \rangle \\[2mm]
U''(t)/2 &= \langle g^{-1}, \Gamma_{\!2}(g,g) \rangle - \langle g^{-2}, \Gamma(g, \Gamma(g,g)) \rangle + \\
& \quad \langle g^{-3}, \Gamma^2(g,g) \rangle.
\end{cases}
$$

Notice that, in this case, the function $U(t)$ is decreasing.

Suppose now that there are 3 positive constants $A$, $B_1$ and $B_2$ such that, for every smooth strictly positive function $g$ on $E$ with $\langle g \rangle = 1$, the following inequality holds:

$$
\left.
\begin{aligned}
&\langle g^{-1}, \Gamma_{\!2}(g,g) \rangle - \langle g^{-2}, \Gamma(f, \Gamma(f,f)) \rangle + \langle g^{-3}, \Gamma^2(g,g) \rangle \geq \\
&A\langle g^{-1}, \Gamma(g,g) \rangle + B_1 \langle g^{-3}, \Gamma^2(g,g) \rangle + B_2 \langle g^{-1}, \Gamma(g,g) \rangle^2.
\end{aligned}
\right\}
\tag{3.6}
$$

If we notice that

$$
\langle g^{-3}, \Gamma^2(g,g) \rangle = \langle g \rangle \langle g^{-3}, \Gamma^2(g,g) \rangle \geq \langle g^{-1}, \Gamma(g,g) \rangle^2,
$$

then we get a differential inequality

$$
U''(t)/2 + AU'(t) \geq BU'^2(t),
\tag{3.7}
$$

with $B = B_1 + B_2$. This last differential inequality leads at once to the majorization

$$
U(0) - U(\infty) \leq \frac{1}{2B} \log\{1 - \frac{B}{A} U'(0)\}.
$$

But we have $U(\infty) = \langle f \rangle \log\langle f \rangle = 0$ and $U'(0) = -\langle f^{-1}, \Gamma(f,f) \rangle$. So, this last inequality becomes

$$
\langle f \rangle = 1 \Rightarrow \langle f \log f \rangle \leq \frac{1}{2B} \log\{1 + \frac{B}{A} \langle f^{-1}, \Gamma(f,f) \rangle\}.
$$

Now, it remains to replace $f$ by $f^2$ in the previous formula to get a (TWS) inequality.

All we have to do now is to check inequality (3.6). For that, we will use the (CD) hypothesis and begin with a rather complicated lemma:

**Lemma 3.3.**—*Suppose that (CD) holds for some constants $n > p$ and $\rho > 0$. Then, consider 8 constants $\alpha$, $\beta$, $\gamma$, $\delta$, $\varepsilon$, $\zeta$, $\eta$, $\theta$ satisfying the following assumptions*

*(1) $\alpha > 0$, $\varepsilon \geq 0$, $\eta \geq 0$, $\alpha + n\gamma \geq 0$, and $\varepsilon \alpha n - (n-1)\beta^2 \geq 0$;*

*(2) $\alpha(\beta + n\delta)^2 \leq (\alpha + n\gamma)[\varepsilon \alpha n - (n-1)\beta^2]$;*

*(3) $\eta[\varepsilon \alpha n - (n-1)\beta^2] \geq \alpha n \zeta^2$.*

*(4) $\theta = \dfrac{-\alpha\zeta(\beta + n\delta)}{\varepsilon \alpha n - (n-1)\beta^2}$.*

*Then, for every positive function $g$ on $E$, and every smooth function $f$, one has*

$$\left. \begin{aligned} &\alpha g^2[\mathbb{L}_2(f,f) - \rho\Gamma(f,f)] - \beta g\Gamma(f,\Gamma(f,f)) + \gamma g^2(\mathbf{L}f)^2 \\ &-2\delta g\mathbf{L}f\Gamma(f,f) - 2\zeta g^2\Gamma(f,f) + \varepsilon\Gamma(f,f)^2 - 2\theta g^3\mathbf{L}f + \eta g^4 \geq 0. \end{aligned} \right\} \tag{3.8}$$

Before going into the proof of this lemma, let us show how we use it. We take $g = f$ and multiply everything by $f^{-3}$. Then, we integrate this expression with respect to the measure $\mu$. We will use the following set of formulae:

$$\left\{ \begin{aligned} \langle \mathbf{L}f \rangle &= 0 \\ \langle f^{q-1}, (\mathbf{L}f)^2 \rangle &= \langle f^{q-1}, \mathbb{L}_2(f,f) \rangle + \frac{3}{2}(q-1)\langle f^{q-2}, \Gamma(f,\Gamma(f,f)) \rangle + \\ (q-1)(q-2)\langle f^{q-3}, \Gamma^2(f,f) \rangle \\ \langle f^{q-2}, \mathbf{L}f\Gamma(f,f) \rangle &= -\langle f^{q-2}, \Gamma(f,\Gamma(f,f)) \rangle - (q-2)\langle f^{q-3}, \Gamma^2(f,f) \rangle. \end{aligned} \right.$$

Using this with $q = 0$, we will get a whole family of inequalities, where only the 4 quantities $\langle f^{-1}, \Gamma(f,f) \rangle$, $\langle f^{-1}, \mathbb{L}_2(f,f) \rangle$, $\langle f^{-2}, \Gamma(f,\Gamma(f,f)) \rangle$, $\langle f^{-3}, \Gamma^2(f,f) \rangle$ will appear. Look carefully at these inequalities: if we set $\eta$ to it's optimal value and replace $\theta$ by it's value, then we get for every real $\zeta$ a positive expression which is quadratic in $\zeta$. Then, if we write that it's discriminant is negative, we get a family of inequalities of the form

$$a\langle f^{-1}, \mathbb{L}_2(f,f) \rangle + b\langle f^{-2}, \Gamma(f,\Gamma(f,f)) \rangle +$$
$$+c\langle f^{-3}, \Gamma^2(f,f) \rangle + d\langle f^{-1}, \Gamma(f,f) \rangle + e\langle f^{-1}, \Gamma(f,f) \rangle^2 \geq 0,$$

where the constants $a$, $b$, $c$, $d$, $e$, depend only on the five parameters $\alpha$, $\beta$, $\gamma$, $\delta$, $\varepsilon$. Then we check the values of these 5 parameters for which we get an inequality (3.6). In the end, we get the inequality (3.7) with $A = \alpha\rho$ and $B = \frac{X(\alpha)}{n}$, where $\alpha$ and $X(\alpha)$ are as in the theorem (3.2).

It remains to show how we get lemma (3.3). We write everything in a local system of coordinates, and we set, with $L = \Delta + \nabla h$, $X = \nabla h$; $Y = \nabla f$, $M = \nabla\nabla f$, $A = \mathrm{Ric}(L)$: $X$ and $Y$ are vectors in the euclidian space $\mathcal{R}^p$; $A$ and $M$ are symmetric $p \times p$ matrices. With these notations, we have $\mathbb{L}_2(f,f) = |M|^2 + {}^tYAY$, $\mathbf{L}f = \mathrm{tr}(M) + Y.X$, $\Gamma(f,f) = |Y|^2$ and $\Gamma(f,\Gamma(f,f)) = 2{}^tYMY$.

Now, the hypothesis (CD) can be written

$$^tYAY \geq \frac{1}{n-p}|Y.X|^2 + \rho|Y|^2.$$

The lemma now will be an immediate consequence of the following:

**Lemma 3.4.**—*With the same conditions on $\alpha$, $\beta$, $\gamma$, $\delta$, $\varepsilon$, $\zeta$, $\eta$, $\theta$ than in lemma (3.3), and for every vector $X$ and $Y$, every symmetric matrix $M$ in $\mathcal{R}^p$, the following inequality holds*

$$\alpha(\|M\|^2 + \tfrac{1}{n-p}(Y.X)^2) - 2\beta\,^tYMY + \gamma(t + Y.X)^2$$
$$-2\delta(t + Y.X)\|Y\|^2 - 2\zeta\|Y\|^2 + \varepsilon\|Y\|^4 - 2\theta(t + Y.X) + \eta \geq 0.$$

The proof of this lemma is purely computationnal and rather boring. We refer the reader to [B].

—**References**

[A] AUBIN (T)— **Non linear analysis on manifolds,** MONGE-AMPÈRE equations , Springer, Berlin-Heidelberg-New-York, 1982.

[B1] BAKRY (D)— La propriété de sous-harmonicité des diffusions dans les variétés, *Séminaire de probabilités XXII*, Lecture Notes in Math. 1321, 1988, Springer, p.1-50.

[B2] BAKRY, D.— Inégalités de SOBOLEV faibles: un critère $\Gamma_2$, to appear in Séminaire de Probablités, 1990.

[BE1] BAKRY (D), EMERY (M)— Diffusions hypercontractives, *Séminaire de probabilités XIX*, Lecture Notes in Math. 1123, 1985, Springer, p.177-206.

[BE2] BAKRY D., EMERY, M.— Hypercontractivité des semigroupes de diffusion, Comptes Rendus Acad. Sc., t.299 , série 1, n° 15, 1984, p.775-778.

[BE3] BAKRY (D), EMERY (M)— Inégalités de SOBOLEV pour un semigroupe symétrique, Comptes Rendus Acad. Sc., t.301 , série 1, n° 8, 1985, p.411-413.

[BE4] BAKRY, D., EMERY M.— Propaganda for $\Gamma_2$, *From local times to global geometry, control and physics*, 1986, K.D. ELWORTHY, Pitmann Research Notes in Mathematics, Longmann Sc& Tech., p.39-46.

[BM]  BAKRY (D) , MICHEL (D)— Inégalités de SOBOLEV et minorations du semigroupe de
      la chaleur, Ann. Fac. Sc. Toulouse, to appear, 1990.

[CS]  CARLEN (E), STROOK (D.W.)— Hypercontractivity, ultracontractivity, SOBOLEV in-
      equalities and all that, preprint.

[CSF] CARLEN (E), KUSUOKA (S), STROOK (D.W.)— Upperbounds for symmetric MARKOV
      transition functions, Ann. Inst. H.POINCARRÉ, vol. 23, 1987, p.245-287.

[D]   DAVIES (E.B.)— **Heat kernels and spectral theory** , Cambridge University Press,
      Berlin-Heidelberg-New-York, 1989.

[DS]  DAVIES (E.B.), SIMON (B.)— Ultracontractivity and the heat kernel for SCHRÖDINGER
      operators and DIRICHLET laplacians, J. Funct. Anal., vol. 59, 1984, p.335-395.

[DSt] DEUSCHEL (J.D.), STROOCK (D.W.) — **Large Deviations** , Ac. Press, Berlin-Heidel-
      berg-New-York, 1989.

[G]   GROSS (L)— Logarithmic SOBOLEV inequalities, Amer. J. Math., vol. 97, 1976, p.1061-
      1083.

[V]   VAROPOULOS (N)— HARDY-LITTLEWOOD theory for semigroups, J. Funct. Anal., vol.
      63, 1985, p.240-260.

# Feynman's Functional Calculus and Stochastic Calculus of Variations

## Ana Bela Cruzeiro and Jean Claude Zambrini

**Key words:** Quantum Mechanics, Stochastic calculus of variations, integration by parts formula, Markov processes, Bernstein processes, Euclidean Quantum Mechanics

## 1. Introduction

The mathematical structure of Quantum Mechanics is usually introduced as a calculus of non-commuting self-adjoint (unbounded) operators, the "observables," on a Hilbert space of "states" (cf. [15]). There is no doubt that Quantum Mechanics is consistent and describes correctly many experiments, but we are supposed to renounce completely the visualization of quantum phenomena in space–time.

Consider, for example, the famous commutation relations $Q_\ell P_m - P_m Q_\ell = i\hbar\delta_{\ell m}$, where $\hbar$ is the Planck constant and $P = -i\hbar\nabla$, $Q = x$ are the momentum and position observables, respectively. These relations are presented as incompatible with the classical concept of trajectories $t \mapsto \omega(t)$ in a path space $\Omega \equiv C([s, u]; \mathbb{R}^3)$, because position and momentum measurements interfere with one another in such a way that both observables cannot have simultaneously a sharp value.

There is nevertheless a space–time approach to non-relativistic Quantum Mechanics due to R. Feynman (1948), which corresponds to a fascinating comeback of classical ideas, in particular the classical calculus of variations. According to Feynman (cf. [6], p. 173), most of the laws of (nonrelativistic) Q. M. follow from the following (formal) *integration by parts formula*:

$$< \delta F[\omega](\delta\omega) >_S = -\frac{i}{\hbar} < F\delta S[\omega](\delta\omega) >_S \qquad (1.1)$$

where $\delta F[\omega](\delta\omega)$ is the directional derivative of a regular functional $F$ of $\omega(\cdot)$, the expectations $< \cdot >_S$ are computed formally with respect to the weight $e^{iS[\omega]/\hbar}\mathcal{D}\omega$, which is used as a measure on the path space $\Omega^y = \{\omega \in C^2([s, t]; \mathbb{R}^3) : \omega(t) = y\}$. $\mathcal{D}\omega = \prod_{s \leq \tau \leq t} d\omega(\tau)$ and $S[\omega]$ is the action functional of the underlying classical system to be quantized, namely $S[\omega] = \int_s^t L(\dot{\omega}(\tau), \omega(\tau))d\tau$, $L$ being its Lagrangian. Starting with Cameron in 1960 [3] there is a depressive number of theorems showing that this cannot make mathematical sense. Nevertheless, let us consider a few formal consequences of this integration by parts formula according to Feynman (cf. [6]), when the Lagrangian is of the simplest form $L(\dot{\omega}, \omega) = \frac{1}{2}|\dot{\omega}|^2 - V(\omega)$:

(i) For $F = 1$, one gets $< \ddot{\omega} >_S = - < \nabla V(\omega) >_S$, a quantum mechanical analogue of the classical Euler–Lagrange equations. In the usual Hilbert space description of the dynamics it corresponds to the Ehrenfest theorem: $\frac{d^2}{d\tau^2} < \chi|Q(\tau)\chi >_2 = - < \chi|\nabla V(Q(\tau))\chi >_2$ for suitable $\chi \in L^2(\mathbb{R}^3)$, where $< | >_2$ stands for the $L^2$ scalar product, $Q(\tau) = e^{i/\hbar H\tau} Q e^{-i/\hbar H\tau}$ (automorphism group of Heisenberg) and $H$ is the energy observable of the system, or Hamiltonian, here $H = -\frac{\hbar^2}{2}\Delta + V$.

(ii) For $F[\omega] = \omega(\tau)$ and an appropriate variation $\delta\omega$, we get, after simplification,

$$< \omega_\ell(\tau)\frac{(\omega(\tau) - \omega(\tau - \Delta\tau))_m}{\Delta\tau} >_S - < \frac{(\omega(\tau + \Delta\tau) - \omega(\tau))_m}{\Delta\tau}\omega_\ell(\tau) >_S$$
$$= i\hbar\delta_{\ell m}$$

This is the space–time version of the Heisenberg commutation relations $Q_\ell P_m - P_m Q_\ell = i\hbar\delta_{\ell m}$. From this, it follows in particular that $< \frac{[\omega(\tau+\Delta\tau)-\omega(\tau)]^2}{\Delta\tau} >_S = i\hbar \mathbb{1}$ ($\mathbb{1}$ is the 3×3 identity matrix), a characteristic of quantum paths, which except for the factor $i$, may seem familiar to a probabilist.

(iii) Taking the same $F$ as in (ii) but another variation, one obtains closed formulas for quantum correlations like $< \omega(t)\omega(\tau) >$, etc.

Feynman translates in this way the main features of Q.M., regarded generally as incompatible with space–time trajectories. As suggested by (ii), however, the relevant trajectories $t \mapsto \omega(t)$ should not be differentiable (what about $\dot{\omega}(t)$ then ?). It is natural to regard Feynman's strategy as an attempt to substitute the non-commutative algebra of operators by a (commutative) stochastic calculus of Markovian diffusions $\omega(t) \equiv z_t(\omega)$. This has never been fully done. One faces at least three kinds of difficulties in implementing Feynman's program:

a) The theory of Markovian diffusions is not time-symmetric as are Classical Mechanics or Quantum Mechanics. The usual version of the Markov property for a stochastic process $z_t, t \in I$, namely $P(z_t \in B|\mathcal{P}_s) = P(z_t \in B|z_s)$, where $\mathcal{P}_s$ is an increasing filtration (the "past"), $s < t \in I$, selects a sense of time, the forward one, given an initial distribution. This refers to the "causa efficiens," insufficient for Quantum Physics, as pointed out repeatedly by Feynman.

b) There is no positive measure (or, equivalently, no stochastic process) like Feynman's one, as already noticed.

c) Functionals of Brownian motion, like, for instance, Itô's stochastic integrals ([9]), are in general not even continuous with respect to the "reasonable" topologies of the Wiener space. Also not all the directions $\delta\omega$ are admissible for derivation. One obviously needs a differential calculus and, in particular, an integration by parts formula, adapted to such irregular functionals.

In the following, we shall describe recent progress in this direction. This can be regarded as a chapter in a program of reinterpretation of Feynman's ideas ("Euclidean Quantum Mechanics"). The general mathematical framework is summarized in another contribution to these proceedings ([10]).

## 2. The Bernstein diffusions

The lack of symmetry in the time of the theory of Markovian diffusions (problem a)) can be fixed by the use, in our framework, of a decreasing filtration $\mathcal{F}_t$ as well as a backward transition probability $P^*(z_t \in B|z_u)$, $u > t$ in $I$, given a final distribution. This filtration $\mathcal{F}_t$ describes a "causa finalis." On the other hand, the transition probabilities of the underlying processes give the evolution (simultaneously backwards and forwards) of the initial and final distributions, according to the laws which come from Quantum Mechanics. This construction has been suggested by Schrödinger's observation that $\frac{d}{dt}\int \eta\eta^*(y,t)dy = 0$, where $\eta^*$ is a (positive) solution of the initial value problem (2.1) and $\eta$ a (positive) solution to the final value problem (2.2).

Although it is possible to consider more general Hamiltonians, let us restrict ourselves to the case where $H = -\frac{\hbar^2}{2}\Delta + V$ is a self-adjoint lower bounded operator in $L^2(\mathbb{R}^3)$, with lower bound $e_0$. Then $e^{-t/\hbar(H-e_0)}$, $t > 0$, is a strongly continuous contraction semigroup. Let $I = [-T/2, T/2]$. Taking $\chi$ and $\chi^*$ positive and bounded in $\mathcal{D}(e^{(T/2)H})$, the set of analytic vectors for $H$, we can solve, in the strong $L^2(\mathbb{R}^3)$ sense, the (forwards) heat equation:

$$\begin{cases} -\hbar\frac{\partial\eta^*}{\partial t} = H\eta^* \\ \eta^*(x,0) = \chi^*(x), \, t \in [-T/2, 0] \end{cases} \tag{2.1}$$

as well as the backwards equation:

$$\begin{cases} \hbar\frac{\partial\eta}{\partial t} = H\eta \\ \eta(x,0) = \chi(x), \, t \in [0, T/2] \end{cases} \tag{2.2}$$

For the discussion of the existence of $\eta$, $\eta^*$, cf[1].

Furthermore, we shall assume the following hypothesis on the potential:

I) $V$ is such that the integral kernel of $e^{-(t-s)H}$ is well-defined, jointly continuous in $x, y$ and $(t - s)$, and strictly positive.

(II) $\sup_x E_x(\exp \alpha \int_0^{T/2} |V|(\hbar^{1/2}w(t))dt) < +\infty$ and

$$\sup_x E_x(\exp \alpha \int_{-T/2}^0 |V|(\hbar^{1/2}w_*(t))dt) < +\infty$$

for some $\alpha > 1$, where $w(t)$ is a $\mathcal{P}_t$-Wiener process and $w_*(t)$ an $\mathcal{F}_t$-Wiener process.

Let $h(t, x, dy)$ denote the integral kernel of $e^{-tH}$ in $L^2(dx)$. The construction described in the second paragraph of [10] yields a Markovian Bernstein process defined in $I = [-T/2, T/2]$ such that, for every $t_0 = -T/2 \leq t_1 \leq \ldots \leq t_n \leq T/2 = t_{n+1}$

$$P(z_{-T/2} \in B_0, z_{t_1} \in B_1, \ldots, z_{t_n} \in B_n, z_{T/2} \in B_{n+1})$$

$$= \int_{B_0 \times B_1 \times} \cdots \int_{\times B_n \times B_{n+1}} \eta^*_{-T/2}(x) \Pi_{i=1}^{n+1} h(t_i - t_{i-1}, x_{i-1}, dx_i) \eta_{T/2}(x_{n+1}) dx$$

This expression also shows that the initial and final distributions are simultaneously propagated forwards and backwards. In fact, another way of looking at these processes is by taking, as initial distribution, $\eta_{-T/2}(x)\eta^*_{-T/2}(x)dx$ whose (forward) evolution is given by the transition probability density (with respect to $dy$)

$$p(s, x, t, y) = h(s, x, t, y)\frac{\eta(y, t)}{\eta(x, s)} \tag{2.3}$$

and, on the other hand, a final distribution $\eta_{T/2}(x)\eta^*_{T/2}(x)dx$, together with the (backward) transition (with respect to $dx$)

$$p^*(s, x, t, y) = \frac{\eta^*(x, s)}{\eta^*(y, t)} h(s, x, t, y). \tag{2.4}$$

It is possible to show ([16]) that these processes are Markovian diffusions, satisfying the following $\mathcal{P}_t$ and $\mathcal{F}_t$ Itô's stochastic differential equations

$$dz_t = \hbar^{1/2} dw_t + \hbar \frac{\nabla \eta}{\eta}(z_t)dt$$

$$d_* z_t = \hbar^{1/2} d_* w_{*t} - \hbar \frac{\nabla \eta^*}{\eta^*}(z_t)dt,$$

where $w_*$ is an $\mathcal{F}_t$-Brownian motion and $d_*$ denotes the backwards differential $d_* f(t) = f(t) - f(t - dt)$ $(dt > 0)$. They have been introduced in 1985-86 under the name of Bernstein diffusions (cf. [16]). Many of their properties are crucial from the quantum dynamical point of view. For instance, notice that since $\eta \leftrightarrow \eta^*$ under time reversal, the theory is time symmetric, as it should be. The fact that, at each $t \in I$, $P(z_t \in B) = \int_B \eta\eta^*(x, t)dx$, can be looked upon as the analogue of the Born interpretation associated with the Schrödinger equation. We also stress the fact that to each potential $v$ is associated a family of stochastic processes, indexed by their initial (and final) distributions.

In the particular situation where we restrict ourselves to time independent (or stationary) states, additional difficulties arise (notably the problem of the "nodes," or zeros of the states) but probabilistic techniques developed by R. Carmona, for example, in another context ([4]) can be used.

From the expressions (2.3) and (2.4) for the transition densities of the process and Feynman–Kac formula, we obtain the following result, valid for any smooth (in particular with non singular drifts) Bernstein diffusion:

**Lemma 2.1.** *Let $z(t)$ be a Bernstein diffusion associated with $H = -\frac{\hbar^2}{2}\Delta + V$, for $V$ satisfying (I), (II). Let $\mu^\hbar$ be the Wiener measure of parameter $\hbar$ and initial distribution given by the law of $z(0)$ on the space of paths $C([0, T/2]; \mathbb{R}^3)$. Then the law of $z$ is absolutely continuous with respect to $\mu^\hbar$ with density $\rho$ given by:*

$$\rho[z] = \frac{\eta(z(T/2), T/2)}{\eta(z(0), 0)} \exp -\frac{1}{\hbar} \int_0^{T/2} V(z(t))dt$$

If $\mu_*^\hbar$ denotes the $\mathcal{F}_t$-Wiener measure of parameter $\hbar$ and final law given by the law of $z(0)$ on the space $C([-T/2, 0]; \mathbb{R}^3)$ then the corresponding density $\rho_*$ of $z_*$ with respect to $\mu_*^\hbar$ is given by:

$$\rho_*[z] = \frac{\eta^*(z(-T/2), -T/2)}{\eta^*(z(0), 0)} \exp -\frac{1}{\hbar} \int_{-T/2}^0 V(z(t))dt$$

Notice that the time reversal of Bernstein diffusions can also be regarded as a transformation of measure, in the following sense:

**Proposition 2.2.** *Let $z(t), t \in [-T/2, T/2]$ be a Bernstein diffusion. Then the law of process $\hat{z}(t) = z(-t)$ is absolutely continuous with respect to the one of $z(t), t \in [s, \tau] \subset [0, T/2]$, with Radon–Nikodym density given by:*

$$\rho_s^\tau[z] = \frac{\eta(z(s), s)}{\eta(z(\tau), \tau)} \frac{\eta^*(z(\tau), -\tau)}{\eta^*(z(s), s)}$$

## 3. Malliavin Calculus

The Malliavin calculus (cf. [12]) is a probabilistic technique which is particularly powerful to prove absolute continuity and smoothness of densities of probability laws of functionals of the Wiener process and has been developed mainly with this application in view. Moreover, it can be regarded as an infinite-dimensional differential calculus adapted to nonregular functionals, a very natural point of view in relation to quantum physics. In this section, we recall some of its basic principles.

Let us denote by $\mathcal{H}$ the Cameron–Martin subspace of $\Omega = \{\omega \in C([0, T/2]; \mathbb{R}^3) : \omega(0) = 0\}$, namely:

$$\mathcal{H} = \{\varphi \in \Omega : \dot{\varphi} \text{ exists and } \int_0^{T/2} \|\dot{\varphi}(s)\|^2 ds < +\infty\}$$

This is a Hilbert space with respect to the scalar product $(\varphi_1|\varphi_2)_1 = \int_0^{T/2} \dot{\varphi}_1(s) \cdot \dot{\varphi}_2(s)ds$. $\Omega$ being considered as a Banach space with the uniform topology, the triple $(\Omega, \mathcal{H}, \mu^\hbar)$ is the so-called classical Wiener space (with parameter $\hbar$). A Wiener functional is a measurable function defined on a Wiener space. Typical examples are solutions of stochastic differential equations, which in general are not continuous.

As is well-known, there is no infinite-dimensional analogue of the Lebesgue measure (like the one involved in (1.1)). Still, by the Cameron–Martin theorem, the measure induced by a translation $\omega \to \omega + \varphi$, with $\varphi \in \mathcal{H}$, is absolutely continuous with respect to the Wiener measure (actually this property characterizes the space $\mathcal{H}$). In the Malliavin Calculus the Cameron–Martin subspace plays the role of a "tangent space." For a Wiener functional $F : \Omega \to \mathcal{G}$, where $\mathcal{G}$ is a Hilbert space, one defines the derivative of $F$ along the direction $\phi \in \mathcal{H}$ as the following (a. s.) limit:

$$\mathbb{D}_\varphi F[\omega] = \lim_{\varepsilon \to 0} \frac{1}{\varepsilon}(F[\omega + \varepsilon\varphi] - F[\omega])$$

The gradient of $F$ is the linear operator $\nabla F : \Omega \to \mathcal{L}(H; \mathcal{G})$ defined by $\nabla F[\omega](\varphi) = \mathbb{D}_\varphi F[\omega]$.

Let $e_1, \ldots, e_k, \ldots$ be an orthonormal basis of $\mathcal{H}$. If $\nabla F$ is a Hilbert–Schmidt operator, namely if $\sum_k \|\mathbb{D}_{e_k} F[\omega]\|_\mathcal{G}^2 < +\infty$ a. s., then, by considering the Hilbert structure of this class of operators, we can define $\nabla^2 F$. By iterating this procedure, one can define $\nabla^i F[\omega]$ as an $i$-linear Hilbert–Schmidt operator on $\mathcal{H}$ and consider the corresponding Sobolev spaces, for $1 \le r, p < +\infty$:

$$W_r^p(\Omega, \mu^\hbar; \mathcal{G}) = \{F \in L^p(\Omega, \mu^\hbar; \mathcal{G}) : E_{\mu\hbar}\|\nabla^i F\|_{H.S}^p < +\infty \ \forall_{1 \le i \le r}\}$$

with $\|\nabla^i F\|_{H.S}^2[\omega] = \sum_{k_1, \ldots, k_i = 1}^{+\infty} \|\nabla^i F[\omega](e_{k_1}, \ldots, e_{k_i})\|_\mathcal{G}^2$.

For $\Phi \in L^2(\Omega, \mu^\hbar; \mathcal{H})$ the divergence of $\Phi$ is defined as the element $\mathcal{D}_{\mu\hbar}\Phi$ in $L^2(\Omega, \mu^\hbar; \mathbb{R})$ such that

$$E_{\mu\hbar}[(\nabla F|\Phi)_1] = E_{\mu\hbar}[F\mathcal{D}_{\mu\hbar}\Phi] \tag{3.1}$$

for every $F \in W_1^2(\Omega, \mu^\hbar; \mathbb{R})$. By [11] we know that, if $\Phi$ is in $W_r^p$ then $\mathcal{D}_{\mu\hbar}\Phi$ exists and belongs to $W_{r-1}^p$. It is an easy consequence of the Cameron–Martin formula that, if $\Phi[\omega]$ is a regular adapted stochastic process, then the divergence coincides with Itô's integral:

$$\mathcal{D}_{\mu\hbar}\Phi[\omega] = \int_0^{T/2} \frac{d}{dt}\Phi[\omega](t)\hbar^{-1}d\omega(t). \tag{3.2}$$

More generally, it can be regarded as a generalization of Itô's integral for non-adapted processes (it coincides with Skorohod's integral, cf. [8]).

For $\rho \in W_1^{1+\varepsilon}(\Omega, \mu^\hbar; \mathbb{R}^+)$, one can prove the following generalization of (3.1) (cf. [5]):

$$E_{\rho\mu\hbar}[(\nabla F | \Phi)_1] = E_{\rho\mu\hbar}[F(\mathcal{D}_{\mu\hbar}\Phi - (\nabla \log \rho | \Phi)_1)] \tag{3.3}$$

This inequality holds for every $\Phi \in W_1^{2q}(\Omega, \mu^\hbar; \mathcal{H})$ and every $F \in W_1^{2q}(\Omega, \mu^\hbar; \mathbb{R})$, where $q$ is the conjugate exponent of $1 + \varepsilon$.

Let $\Omega_* = \{\omega \in C([-T/2, 0]; \mathbb{R}^3) : \omega(0) = 0\}$. A completely analogous structure can be associated to the Wiener space $(\Omega_*, \mathcal{H}_*, \mu_*^\hbar)$, where $\mu_*^\hbar$ is the Wiener measure associated to the $\mathcal{F}_t$-Brownian motion $w_*$. We remark that most of the results of Malliavin calculus are independent of the filtration. In particular, a non-anticipative stochastic calculus has been developed recently in this framework (cf. [13]) and references within). Still, since Bernstein diffusions are symmetric in time, such calculus is not necessary in our context.

## 4. An integration by parts formula

When applied to Bernstein diffusions, Eq. (3.3) provides an integration by parts formula which can be regarded as a rigorous version of Feynman's formula (1.1) in Quantum Mechanics. For this, we need the following:

**Lemma 4.1.** *Let* $0 < \varepsilon < 1$ *be such that* $\alpha = \max\left(\frac{2}{\hbar}(1 + \varepsilon), \frac{1+\varepsilon}{2\hbar\varepsilon}\right) > 1$. *If the hypothesis (II) on* $V$ *is satisfied for* $\alpha$ *and if* $\int |\nabla V|^{2(1+\varepsilon/1-\varepsilon)} dx < +\infty$, *then the function* $\rho$ *of Lemma 2.1 belongs to the Sobolev space* $W_1^{1+\varepsilon}(\Omega, \mu^\hbar; \mathbb{R})(*)$

**Proof.** The initial distribution of the Bernstein diffusion being given by a bounded positive function $(\chi \chi^*(x)dx)$, we can restrict ourselves to the case $z(0) = x$.

By the hypothesis on $V$ and since $\eta(x, 0)$ is a bounded function, the Feynman–Kac representation formula implies that $\sup_x E_x \eta^{2(1+\varepsilon)}(z(T/2),$ $T/2) < +\infty$; therefore $E_x \rho^{1+\varepsilon} < +\infty$.

We have the following expression for the gradient of $\rho$:

$$\nabla \rho[z] = \nabla \eta(z(T/2), T/2) \, \nabla z(T/2) \exp -\frac{1}{\hbar} \int_0^{T/2} V(z(t))dt$$

$$-\frac{1}{\hbar} \eta(z(T/2), T/2) \exp\left(-\frac{1}{\hbar} \int_0^{T/2} V(z(t))dt\right) \int_0^{T/2} \nabla V(z(t)) \nabla z(t)dt.$$

---

(*) We shall denote by $(\Omega, \mu^\hbar)$ the Wiener space independently of the initial condition of the processes.

Therefore, using Hölder's inequality,

$$E_x\|\nabla\rho\|^{1+\varepsilon} \leq \left(\frac{T}{2}\right)^{1/2} (E_x|\nabla\eta(z(T/2),T/2)|^2)^{(1+\varepsilon)/2}$$

$$\left(E_x \exp\frac{2}{\hbar}\left(\frac{1+\varepsilon}{1-\varepsilon}\right)\int_0^{T/2} V(z(t))dt\right)^{(1-\varepsilon)/2}$$

$$+\frac{1}{\hbar}(E_x\eta^{(1+\varepsilon)/2\varepsilon}(z(T/2),T/2))^{2\varepsilon/(1+\varepsilon)}$$

$$\left(E_x \exp\frac{2}{\hbar}\left(\frac{1+\varepsilon}{1-\varepsilon}\right)\int_0^{T/2} V(z(t))dt\right)^{(1-\varepsilon)/2(1+\varepsilon)}$$

$$\left(E_x \left(\int_0^{T/2} t^{1/2}|\nabla V|z(t)dt\right)^{2[(1+\varepsilon)/(1-\varepsilon)]}\right)^{(1-\varepsilon)/2(1+\varepsilon)}$$

The restriction we have imposed on $V$ implies then that $E_x\|\nabla\rho\|^{1+\varepsilon} < +\infty$.
□

We are now able to prove the integration by parts formula for the Bernstein diffusions:

**Theorem 4.2.** *For every $\mathcal{P}_t$-adapted functional $\Phi \in W_1^{2q}(\Omega,\mu^\hbar;\mathcal{H})$ and every $F \in W_1^{2q}(\Omega,\mu^\hbar;\mathbb{R})$, where $q$ is the conjugate exponent of $1+\varepsilon$, we have:*

$$\hbar E_{\rho\mu\hbar}((\nabla F[z]|\Phi)_1)$$
$$= E_{\rho\mu^\hbar}\left[F[z]\left(\int_0^{T/2}\frac{d}{dt}\Phi[z](t)dz(t) + \int_0^{T/2}\Phi[z](t)\nabla V(z(t))dt\right.\right.$$
$$\left.\left.-\hbar\Phi[z](T/2)\frac{\nabla\eta}{\eta}(z(T/2),T/2) + \hbar\Phi[z](0)\frac{\nabla\eta}{\eta}(x,0)\right)\right]$$

The proof is a consequence of (3.2), (3.3) and the expression for $\rho$ (cf. [5] for details). We notice that a similar formula holds with respect to the backward filtration, on the space $(\Omega_*,\mathcal{H}_*,\rho_*\mu_*^\hbar)$.

## 5. The equations of motion

Let us introduce the following notions of forward and backward derivatives, natural in our time symmetric context:

$$Df(z(t),t) = \lim_{\Delta t\downarrow 0}\frac{1}{\Delta t}E[f(z(t+\Delta t),t+\Delta t) - f(z(t),t)|\mathcal{P}t] \quad \text{and}$$

$$D_*f(z(t),t) = \lim_{\Delta t\downarrow 0}\frac{1}{\Delta t}E[f(z(t),t) - f(z(t-\Delta t),t-\Delta t)|\mathcal{F}_t]$$

By applying Theorem 4.2 to $F = 1$, we can deduce that the Bernstein diffusions starting from $x$ satisfy the (forward) Euler–Lagrange equation (cf. [5] for details):

$$\begin{cases} DDz(t) & = \nabla V(z(t)) \\ z(0) & = x \\ Dz(T/2) & = \hbar\frac{\nabla\eta}{\eta}(z(T/2), T/2) \end{cases} \quad \text{a.s. for } t \in [0, T/2] \qquad (5.1)$$

With respect to the backward filtration (and using Malliavin calculus on the space $(\Omega_*, \mathcal{H}_*, \rho_*\mu_*^\hbar)$), we similarly obtain the (backward) Euler–Lagrange equation:

$$\begin{cases} D_*D_*z(t) & = \nabla V(z(t)) \\ z(0) & = x \\ D_*z(-T/2) & = -\hbar\frac{\nabla\eta^*}{\eta^*}(z(-T/2), -T/2) \end{cases} \quad \text{a.s for } t \in [-T/2, 0] \qquad (5.2)$$

Actually, the validity of (5.2) follows from its forward version (5.1) and the time symmetry of Bernstein measures.

One may ask whether there are uniqueness results for solutions of equations (5.1) and (5.2) and in what sense this problem can be formulated. Consider a process given by $dz_t = \hbar^{1/2}dw_t + b(z_t, t)dt$, $z(0) = x$, where the diffusion coefficient is fixed and $b$ is regular enough so that the law of $z_t$ is absolutely continuous with respect to Wiener measure. Equation $DDz = \nabla V$ can be written, using Itô's formula, as $\left(\frac{\partial}{\partial t}b + \frac{\hbar}{2}\Delta b + b.\nabla b\right)(z(t), t) = \nabla V(z(t))$, where $(b.\nabla b)^i = b.\nabla b^i$. Since the boundary conditions of Eq. (5.1) imply that $b$ is a gradient, $b = \nabla I$, it can easily be transformed into a Hamilton–Jacobi equation

$$\left(\frac{\partial}{\partial t}I + \frac{\hbar}{2}\Delta I + \frac{1}{2}\|\nabla I\|^2\right)(x, t) = V(x) + f(t)$$

(where $f$ is arbitrary) with a final Cauchy data. This equation is familiar in Optimal Stochastic Control Theory (cf. [7]) and has a unique solution.

One may also look for a direct proof of uniqueness: suppose we are given two diffusion processes solving a stochastic differential equation with diffusion coefficient $\hbar\mathbb{1}$, namely $z$ and $\tilde{z}$ such that $DDz = F(z) = DD\tilde{z}, z(0) = \tilde{z}(0), Dz(T/2) = D\tilde{z}(T/2)$. Furthermore, assume that the following supplementary conditions hold: $F$ is a Lipschitz function and $E\int_0^{T/2}\|\nabla Dz(t)\|^2dt$ as well as $E\int_0^{T/2}\|\nabla D\tilde{z}(t)\|^2dt$ are finite. Then

$$E\|z(t) - \tilde{z}(t)\|^2 = E\left\|\int_0^t [Dz(s) - D\tilde{z}(s)]ds\right\|^2 \leq 2\hbar E\left\|\int_t^{T/2} [\nabla Dz(s)\right.$$

$$\left. - \nabla D\tilde{z}(s)]dw_s dt\right\|^2 + 2E\left\|\int_0^t\int_t^{T/2} [DDz(s)\right.$$

$$\left. - DD\tilde{z}(s)]dsdt\right\|^2 \leq 2\hbar E\int_t^{T/2}\|\nabla Dz(s) - \nabla D\tilde{z}(s)\|^2ds$$

$$+ 2\left(\int_0^t\int_t^{T/2}\|F(z(s)) - F(\tilde{z}(s))\|dsdt\right)^2$$

Using the assumptions on $z$ and $\tilde{z}$, Gronwall's lemma implies that $z(t) = \tilde{z}(t)$ a.e., for $t \in [0, T/2]$. This method seems to have the disadvantage of restricting too much the allowed class of processes.

## 6. Variational principles

As in classical calculus of variations, the solutions of Euler–Lagrange equations (5.1) and (5.2) correspond to extremals of the associated action functionals. Let $z$ be a process given by $dz_t = \hbar^{1/2} dw_t + b(z_t, t) dt$, $z(0) = x$, and whose law is absolutely continuous with respect to the Wiener measure, namely is of the form $\rho \mu^\hbar$ for some density $\rho > 0$. Let $F$ be a functional defined on $(\Omega, \rho \mu^\hbar)$. We say that $z$ is an extremal for $F$ (on a set of diffusions $d\tilde{z}_t = \hbar^{1/2} dw_t + \tilde{b}(\tilde{z}_t, t) dt$, $\tilde{z}(0) = x$ having the law abs. cont. w. r. t. $\mu$) if $(\nabla F | \varphi)_1 = 0$ a. s. for every $\varphi \in \mathcal{H}$.

Let us consider action functionals of the form

$$F[z] = E_{0,x} \int_0^{T/2} L(Dz, z, s) ds - E_{0,x} I(z(T/2), T/2) \qquad (6.1)$$

where $L(\dot{\omega}, \omega, s)$ is the Euclidean Lagrangian of the underlying classical system. We have the following generalization of Theorem 5.3 of [5]:

**Theorem 6.1.** *Assume that the functions $L$ and $I$ in (6.1) are of class $C^1$ and $I$ satisfies $\nabla I(z(T/2), T/2) = \frac{\partial L}{\partial Dz}(Dz(T/2), z(T/2), T/2)$. Then $z$ is an extremal for $F$ iff the following Euler–Lagrange equation holds:*

$$D \frac{\partial L}{\partial Dz} - \frac{\partial L}{\partial z} = 0 \qquad a.s.$$

**Proof.** We first notice that the condition $(\nabla F | \varphi)_1 = 0$, $\forall \varphi \in \mathcal{H}$, implies that $F$ is constant a. e. in $(\Omega, \rho \mu^\hbar)$ (cf., for example, [14]). Therefore if $\Phi : \Omega \to \mathcal{H}$ is "regular" enough (so that the measure induced by $z + \Phi[z]$ is abs. cont. w. r. t. $\rho \mu^\hbar$), we have $(\nabla F[z] | \Phi[z])_1 = 0$. Now, using the expression (6.1) for $F$:

$$(\nabla F | \Phi)_1 = E_{0,x} \int_0^{T/2} \left( \frac{\partial L}{\partial z} \Phi + \frac{\partial L}{\partial Dz} D\Phi \right) ds$$

$$- E_{0,x} \nabla I(z(T/2), T/2) \Phi[z](T/2)$$

$$= E_{0,x} \int_0^{T/2} \Phi \left( \frac{\partial L}{\partial z} - D \frac{\partial L}{\partial Dz} \right) ds + E_{0,x} \Phi[z](T/2)$$

$$\left( \frac{\partial L}{\partial Dz}(z(T/2), T/2) - \nabla I(z(T/2), T/2) \right)$$

by Itô calculus ([9]). Finally, from

$$(\nabla F|\Phi)_1 = E_{0,x} \int_0^{T/2} \Phi \left( \frac{\partial L}{\partial z} - D\frac{\partial L}{\partial Dz} \right) ds = 0$$

for arbitrary $\Phi$, the Euler–Lagrange equations follow. □

In a Quantum-Mechanical framework, the equations (5.1) are the Euler–Lagrange equations associated to the following (*classical*, but regularized) Euclidean action functional:

$$F[z] = E_{0,x} \int_0^{T/2} \left( \frac{1}{2}\|Dz(s)\|^2 + V(z(s))\| \right) ds - \hbar E_0 \log \eta(z(T/2), T/2)$$
(6.2)

Notice the "renormalization" of Feynman's time derivative involved here.

Corresponding results also hold for the backward Euler–Lagrange equations.

It is a familiar result in classical calculus of variations (cf., for ex. [2]) that a necessary condition for a trajectory $\omega$ in $C^1([0, T/2])$ to be a (local) minimum for the "classical limit" of the action (6.1), namely the Boltz action:

$$F[\omega] = \int_0^{T/2} L(\dot{\omega}, \omega, s)ds - I(\omega(T/2))$$
(6.1')

is that the Euler–Lagrange equation

$$\frac{d}{ds}\frac{\partial L}{\partial \dot{\omega}} - \frac{\partial L}{\partial \omega} = 0$$

holds, together with the transversality condition

$$\frac{\partial L}{\partial \dot{\omega}}(\dot{\omega}(T/2), \omega(T/2), T/2) = \nabla I(\omega(T/2))$$

In this sense, Theorem 6.1 is a natural extension of the classical result. It is part of the program of Euclidean Quantum Mechanics to explore systematically such extensions in terms of Malliavin calculus. The Euler–Lagrange equations (5.1), (5.2) are, of course, interpreted as stronger versions of Feynman's dynamical law (i), Section 1.

## 7. Applications

If we apply the formula given by Theorem 4.2 to $F[z] = z(t)$ and

$$\Phi[z](\tau) = \begin{cases} 0 & , \tau < t - \Delta t \\ 1 + \frac{\tau - t}{\Delta t} & , t - \Delta t \le \tau \le t \\ 1 & , \tau > t \end{cases}$$

(in one dimension, for simplicity), we get when $\Delta t \to 0$,

$$\hbar = E[z(t)(D_*z(t) - Dz(T/2) + \int_t^{T/2} \nabla V(z(\tau))d\tau)]$$

and, by an analogous argument on the space $(\Omega_*, \mathcal{H}_*, \rho_* \mu_*^\hbar)$,

$$0 = E[z(t)(Dz(t) - Dz(T/2) + \int_t^{T/2} \nabla V(z(\tau))d\tau)]$$

Therefore, by subtracting the two equations, $E[z(t)D_*z(t) - z(t)Dz(t)] = \hbar$. In three dimensions this yields the Euclidean version of Feynman commutation relations (ii), Section 1 (cf. [10] for another more general derivation of this result):

$$E[z^i(t)D_*z^j(t) - z^i(t)Dz^j(t)] = \hbar\delta_{ij}$$

This relation may be regarded as the strongest possible justification for the existence of two distinct concepts of regularized derivatives in the present probabilistic approach to quantum dynamics.

For    $F[z] = z(t)$    and

$$\phi[z](s) = \begin{cases} s - \tau &, s > \tau \\ 0 &, s \leq \tau \end{cases} \qquad , \text{where} \quad 0 < \tau < T/2,$$

We obtain (again in dimension one for simplicity) the following formulae for the correlation functions:

$$\frac{d}{d\tau}E[z(t)z(\tau)] = -E\left[z(t)\left(\int_\tau^{T/2} \nabla V(z(s)ds + Dz(T/2))\right)\right] \quad \text{for } t < \tau,$$

and

$$\frac{d}{d\tau}E[z(t)z(\tau)] = -E\left[z(t)\left(\int_\tau^{T/2} \nabla V(z(s)ds + Dz(T/2))\right)\right] + \hbar, \text{ for } t > \tau.$$

Another method of studying correlation functions is by means of characteristic functionals. For $f \in \mathcal{H}$, let

$$v(s, x, f) = E_{s,x}\left[\exp i \int_s^{T/2} \frac{d}{d\tau}f(\tau)dz(\tau)\right]$$

$$= E_{s,x}\left[\exp\left(i \int_s^{T/2} \frac{d}{d\tau}f(\tau)dw(\tau)\right)\rho(w)\right]$$

where $\rho$ is the density of the Bernstein process with respect to Wiener measure ($\hbar = 1$). For any $\varphi \in \mathcal{H}$,

$$D_\varphi v(s,x,f)|_{f=0} = iE_{s,x}\left(\int_s^{T/2} \frac{d}{d\tau}\varphi(\tau)dz(\tau)\right)$$

By choosing $\varphi(\tau) = \mathbb{1}_{(s,t)} \cdot \tau$, we obtain:

$$E_{s,x}z(t) = x - iD_\varphi v(s,x,f)|_{f=0}$$

The second derivative of $v$ yields:

$$E_{s,x}((z(t)-x)(z(\tau)-x)) = -D^2_{\varphi\psi}v(s,x,f)|_{f=0},$$

where $\psi(\tau) = \mathbb{1}_{(s,\tau)} \cdot \mathbf{r}$.

These formulas can be recognized as well-defined (Euclidean) versions of Feynman's ones (cf. [6], paragraph 7-4).

## REFERENCES

[1] Albeverio S., Yasue K. and Zambrini J. C., *Euclidean Quantum Mechanics: Analytic Approach*, Ann. Inst. Henri Poincaré, Phys. Th., 49, (3), 259 (1989).

[2] Alexéev V., Tikhomirov V. and Fomine S., *Commande optimale*, Ed. Mir, Moscou (1982).

[3] Cameron R. H., *A family of integrals serving to connect the Wiener and Feynman integrals*, J. Math. Phys., **39**(126), (1960).

[4] Carmona R., *Probabilistic construction of Nelson processes*. In: Proc. Tan. Intern. Symp. Katata and Kyoto, 1985, K. Itô and N. Ikeda ed.

[5] Cruzeiro A. B. and Zambrini J. C., *Malliavin Calculus and Euclidean Quantum Mechanics*, I - Functional Calculus, J. Funct. Anal. , **96**(1), 62–95 (1991).

[6] Feynman R. P. and Hibbs A. R., *Quantum Mechanics and Path Integrals*, McGraw–Hill, New York (1965).

[7] Fleming W. H., *Controlled Markov Processes and Viscosity Solution of Nonlinear Evolution Equations*, Scuola Normale Superiore, Pisa (1986).

[8] Gaveau B. and Trauber P., *L'intégrale stochastique comme opérateur de divergence dans l'espace fonctionnel*, J. Funct. Anal. **46**(230), (1982).

[9] Ikeda N. and Watanabe S., *Stochastic differential equations and diffusion processes*, Second Edition, North-Holland and Kodansha, Amsterdam (1989).

[10] Kolsrud T. and Zambrini J. C., *The general mathematical framework of Euclidean Quantum Mechanics, an outline*, in this volume.

[11] Krée M. and Krée P., *Continuité de la divergence dans les espaces de Sobolev relatifs à l'espace de Wiener*, C. R. Acad. Sci. Paris, t 296, Ser.I, (833),(1983).

[12] Malliavin P., *Stochastic calculus of variations and hypoelliptic operators*, In: Proc. Symp. Stoch. Diff. Eqs., Itô K. ed., Kinokuniya–Wiley, Tokyo (1978).

[13] Nualart D. and Pardoux E., *Stochastic Differential Equations with boundary conditions*, in this book.

[14] Shigekawa I., *Derivatives of Wiener functionals and absolute continuity of induced measures*, J. Math. Kyoto Univ., **20-2**(263), (1980).

[15] Von Neumann J., *Mathematical Foundations of Quantum Mechanics*, Princeton Univ. Press, Princeton (1955).

[16] Zambrini J. C., *New probabilistic approach to the classical heat equation*, In: Proc. of Swansea Meeting, Ed. A. Truman, I. M. Davies, Lect. Notes in Math., **1325**, (205) Springer (1988).

A.B. Cruzeiro
Instituto de Física e Matemática
Av. Prof. Gama Pinto 2
1699 Lisboa Codex, Portugal
and
J.C. Zambrini
Department of Mathematics,
Royal Institute of Technology,
S-10044 Stockholm, Sweden

# Manifolds and Graphs with Mostly Positive Curvatures

## K.D. ELWORTHY*

*Keywords: infinite graphs, Laplacians, spectrum, Riemannian manifolds, ends, Ricci curvature, $\Gamma_2$.*

**Abstract:** Bakry–Emery theory is used to define a 'Ricci curvature' for graphs. An upper bound for the spectral abcissa of the Laplacian of graphs with more than one end is given in terms of this quantity. This is similar to an existing result for manifolds, and the proof of that is also given.

## §0. Introduction

One of the first differential geometric questions to which probabilistic methods were applied was to the possible extensions of Bochner's vanishing theorem. This gives topological restrictions on the possibility of a compact manifold admitting a metric with positive curvature. The main technique was the use of heat equation methods exploited a long time previously by Milgram and Rosenbloom in the context of Hodge theory and in vogue as a tool in topology, for example in Patodi's proof of the Gauss–Bonnet theorem (for references and details see [17], [14]). The idea in this approach to vanishing theorems was to use a probabilistic solution to the heat equation for differential forms to obtain a semigroup domination for that heat semigroup in terms of the heat semigroup for functions and the curvature of the manifold. From this it is clear that a certain positivity of the curvature implies that there are no nonzero harmonic forms, and the topological consequences follow from Hodge theory. Some aspects of this are described in slightly more detail below.

---

*Support was received from NATO Collaborative Research Grants Prog. 0232/87 and the EEC Stimulation Action Plan.

In fact it was in Itô's 1962 International Congress article [20] that the idea of using probabilistic methods was suggested. A lot of work was done in the 70's by Malliavin and co-workers, e.g., [1], [21], [22], [25], [6], and [23] for manifolds with boundary. Semigroup domination was rediscovered in this context and applied to manifolds with boundary by Donnelly and Li [11], see [5] for a useful survey and more references.

In a series of articles the author and S. Rosenberg have followed up these ideas, based on the probabilistic approach, although finally the bulk of the proofs have been analytical with a substantial input of algebraic topology. There have been two main thrusts to that programme: (i) to weaken the conditions on the positivity of the curvature, i.e., to see how much negativity can be allowed and (ii) to obtain conditions for noncompact manifolds on the topology, or geometry, of the manifold 'near infinity', and to apply these to covering spaces of compact manifolds to get new results for nonsimply connected compact manifolds. Such results are described in [16], with a discussion of quantitative behavior in [15]. The article [13] is more introductory, and some of the proofs about connectedness at infinity can be improved as mentioned in [15], and as described below.

The curvature which arises in the semigroup domination for 1-forms is the Ricci curvature. At a meeting in Oberwolfäch, D. Bakry pointed out that the abstract approach to Ricci curvature for suitable operators, via "$\Gamma_2$", should give analogous results in different situations. A computation showed that that was indeed the case for the Laplace operator for certain infinite graphs. This is described in the next section. For comparison the more standard case of manifolds is then presented, using the approach outlined in [15].

## §1. Graphs

**A.** A *graph* $G$ will consist of a set $G_0$ together with a symmetric relation $G_1$, the relation of being *neighbors*. Thus $G_1$ is a subset of $G_0 \times G_0$. It is the set of directed edges and is taken not to intersect the diagonal. We write $x \sim y$ if $(x, y) \in G_1$, and $x \in N(y)$ if $x \sim y$. Often $[x, y]$ will be used instead of $(x, y)$ for elements of $G_1$.

Following the weighting used in [9], define

$$m(x) = \#N(x) \qquad x \in G_0.$$

This will be assumed finite for each $x$. Set

$$\ell_0^2(G) = \left\{ f : G_0 \to \mathbb{R} : \sum_{G_0} m(x) f(x)^2 < \infty \right\}$$

$$\ell_1^2(G) = \left\{ \varphi \colon G_1 \to \mathbb{R} \colon \varphi([x,y]) = -\varphi([y,x]) \text{ and } \sum_{G_1} \varphi([x,y])^2 < \infty \right\}.$$

These correspond to functions and '1-forms' respectively. They become Hilbert spaces with the obvious inner product for functions and using the sum over undirected edges $G_1'$

$$\langle \varphi_1, \varphi_2 \rangle = \sum_{G_1'} \varphi_1([x,y]) \varphi_2([x,y])$$

for the 1-forms. (In other words you arbitrarily choose one of $[x,y]$ or $[y,x]$ for the sum and forget the other.) There is the differentiation

$$d \colon \ell_0^2 \to \ell_1^2$$

given by

$$df([x,y]) = f(y) - f(x)$$

which is a bounded operator with

$$\|d\| \le \sqrt{2}.$$

(This was the reason for using these weightings in [9].) There is the adjoint

$$d^* \colon \ell_1^2 \to \ell_0^2$$

and 'Laplacian' [9]

$$\Delta = -d^* d \colon \ell_0^2 \to \ell_0^2$$

which are calculated to be given by

$$d^* \varphi(x) = \frac{1}{m(x)} \sum_{y \sim x} \varphi([y,x])$$

and

$$\Delta f(x) = \frac{1}{m(x)} \sum_{y \sim x} (f(y) - f(x)).$$

Thus $\Delta$ is a nonpositive bounded self-adjoint operator with $\|\Delta\| \le 2$. Let $\lambda_0(G)$ be the top of its spectrum, so $\lambda_0(G) \le 0$ and

$$\lambda_0(G) = \sup\{\langle \Delta f, f \rangle \colon \|f\| = 1\}$$
$$= \sup\{-\|df\|^2 \colon \|f\| = 1\}$$
$$\ge -2.$$

The formula for $\Delta f$ shows that the Laplacian extends to a bounded linear operator $\Delta: \ell_0^\infty \to \ell_0^\infty$ where $\ell_0^\infty$ is the space of bounded maps $f: G_0 \to \mathbb{R}$ with supremum norm. We will consider the semigroup

$$P_t = e^{\frac{1}{2}\Delta t}$$

acting on $\ell_0^\infty$ or its restriction to $\ell_0^2$ or intermediate spaces $\ell_0^p$, $2 \leq p \leq \infty$ as is convenient.

**B.** In [9] Dodziuk and Kendall gave the analogue of Cheeger's inequality in this context:

$$\lambda_0(G) \leq -\frac{1}{2}\alpha(G)^2$$

where $\alpha(G)$ is the isoperimetric constant

$$\alpha(G) = \inf\left\{ \frac{L(\partial K)}{A(K)} : K \text{ is a finite subgraph of } G \right\}$$

where

$$A(K) = \sum_{x \in K} m(x)$$

is the 'area' of the subgraph $K$ and $L(\partial K)$ is the number of edges of $G$ outward pointing from $K$:

$$L(\partial K) = \#\{[x,y] \in G_1 \text{ s.t. } x \in K, \ y \notin K\}.$$

In fact for a finite subgraph $K$ if $f: G \to \mathbb{R}$ is the characteristic function of the vertices of $K$ then

$$\lambda_0(G) \geq -\|df\|^2/\|f\|^2 = -L(\partial K)/A(K).$$

Thus

$$-\alpha(G) \leq \lambda_0(G) \leq -\frac{1}{2}\alpha(G)^2.$$

For a discussion of $\alpha(G)$ see [18] and the references there, but note the difference in the definition of $\alpha(G)$.

In particular if $G$ is the graph obtained from the 1-dimensional skeleton of a triangulation of the plane with $m(x) \geq 7$ for every vertex $x$ then [9]

$$\alpha(G) \geq \frac{1}{78}.$$

**C.** The following was worked out with D. Bakry:

**Theorem I.** *Suppose G is connected and can be disconnected into two infinite parts by the removal of a finite set of points. Then*

$$\lambda_0(G) \geq \frac{4}{\sup m(x)} - 2.$$

Before giving the proof we will interpret the right-hand side of the inequality in terms of the generalized Ricci curvature of Bakry and Emery and prove a special case of one of their general results.

For $f, g: G_0 \rightarrow \mathbb{R}$ define

$$\Gamma(f, g) = \frac{1}{2}\{\Delta(fg) - f\Delta g - g\Delta f\}$$

and

$$\Gamma_2(f, g) = \frac{1}{2}\{\Delta\Gamma(f, g) - \Gamma(f, \Delta g) - \Gamma(g, \Delta f)\}$$

where the Laplacian has been extended to act on arbitrary functions on $G_0$.

**Lemma 1.** *For $f: G_0 \rightarrow \mathbb{R}$ and $x \in G_0$*

$$\Gamma_2(f, f)(x) \geq \inf_{y \sim x}\left(\frac{2}{m(y)} - 1\right)\Gamma(f, f)(x) + \frac{1}{2}(\Delta f(x))^2.$$

**Proof.** First note that

$$2\Gamma(f, f)(x) = \frac{1}{m(x)}\sum_{y \sim x}(f(y) - f(x))^2.$$

Therefore

$$2\Delta\Gamma(f, f)(x) = \frac{1}{m(x)}\sum_{y \sim x}\left\{\frac{1}{m(y)}\sum_{z \sim y}(f(z) - f(y))^2\right.$$

$$\left. - \frac{1}{m(x)}\sum_{y' \sim x}(f(y') - f(x))^2\right\}$$

$$= \frac{1}{m(x)}\sum_{y \sim x}\frac{1}{m(y)}\sum_{z \sim y}(f(z) - f(y))^2$$

$$- \frac{1}{m(x)}\sum_{y \sim x}(f(y) - f(x))^2$$

$$= \frac{1}{m(x)}\sum_{y \sim x}\frac{1}{m(y)}\sum_{z \sim y}\{(f(z) - f(y))^2 - (f(y) - f(x))^2\}$$

and also

$$2\Gamma(f, \Delta f)(x) = \frac{1}{m(x)} \sum_{y \sim x} (f(y) - f(x))(\Delta f(y) - \Delta f(x))$$

$$= \frac{1}{m(x)} \sum_{y \sim x} (f(y) - f(x)) \left\{ \frac{1}{m(y)} \sum_{z \sim y} (f(z) - f(y)) \right.$$

$$\left. - \frac{1}{m(x)} \sum_{y' \sim x} (f(y') - f(x)) \right\}$$

$$= \frac{1}{m(x)} \sum_{y \sim x} (f(y) - f(x)) \frac{1}{m(y)} \sum_{z \sim y} (f(z) - f(y))$$

$$- \frac{1}{m(x)^2} \left( \sum_{y \sim x} (f(y) - f(x)) \right)^2 .$$

Thus

$$4\Gamma_2(f, f)(x) = 2\Delta\Gamma(f, f)(x) - 4\Gamma(f, \Delta f)(x)$$

$$= \frac{1}{m(x)} \sum_{y \sim x} \left\{ \frac{1}{m(y)} \sum_{z \sim y} ((f(z) - f(y)) - (f(y) - f(x))^2 \right\}$$

$$- \frac{2}{m(x)} \sum_{y \sim x} (f(y) - f(x))^2 + \frac{2}{m(x)^2} \left( \sum_{y \sim x} (f(y) - f(x)) \right)^2$$

$$\geq \frac{1}{m(x)} \sum_{y \sim x} \frac{1}{m(y)} 4 (f(x) - f(y))^2$$

$$- 4\Gamma(f, f)(x) + 2 (\Delta f(x))^2$$

by taking $z = x$ in the sum over $z$ and ignoring any other possible $z$. $\quad\square$

Constants $(n, C)$ are said, [4], to form an *admissible couple* (*dimension, curvature*) for $\Delta$ if $n \geq 1$ and for all $f$

$$\Gamma_2(f, f) \geq C\Gamma(f, f) + \frac{1}{n} (\Delta f)^2.$$

From the lemma we can take $n = 2$ and $C = \inf\{\frac{2}{m(x)} - 1 : x \in G_0\}$. The philosophy of the Bakry–Emery theory is that

$$\rho(x) := \inf_{y \sim x} \left\{ \frac{2}{m(y)} - 1 \right\}$$

is in some sense a lower bound for the 'Ricci curvature' of $G$ at $x$ and $n$ is an upper bound for the 'dimension.' This is by analogy with the corresponding inequality using the Laplace–Beltrami operator on a smooth manifold, [2], see also [3].

The following is a special case of a more general result in the theory.

**Lemma 2.** *For $\rho$ as above*

$$\frac{d}{dt}\Gamma(P_t f, P_t f)(x) \le \left(\frac{1}{2}\Delta - \rho(x)\right)\Gamma(P_t f, P_t f)(x)$$

*for all $x \in G_0$ and $f \in \ell_0^\infty$.*

**Proof.**

$$\frac{d}{dt}\Gamma(P_t f, P_t f) = 2\Gamma\left(\frac{1}{2}\Delta P_t f, P_t f\right)$$

$$= \frac{1}{2}\Delta\Gamma(P_t f, P_t f) - \Gamma_2(P_t f, P_t f) \le \left(\frac{1}{2}\Delta - \rho\right)\Gamma(P_t f, P_t f)$$

by the lemma.   □

For $\varphi : G_1 \to \mathbb{R}$ and $x \in G_0$, set

$$|\varphi|_x = \sqrt{\left\{\frac{1}{m(x)}\sum_{y \sim x}\varphi([x, y])^2\right\}}$$

so that $\Gamma(f, f) = \frac{1}{2}|df|_x^2$ as in the proof of Lemma 1. Note however that

$$\|df\|^2 = \frac{1}{2}\|(|df|)\|^2.$$

**Proof of Theorem I.** Let $K$ be a finite set of points of $G_0$ such that $G_0 - K$ is the union of disjoint sets $A, B$ with no $[x, y]$ in $G_1$ having $x \in A$ and $y \in B$. Take $f : G_0 \to \mathbb{R}$ defined by

$$f(x) = 1 \qquad x \in A$$
$$= 0 \qquad x \in K$$
$$= -1 \qquad x \in B.$$

Since all the operators are bounded on $\ell_0^\infty$

$$dP_t f = df + \int_0^t dP_s \Delta f \, ds. \tag{1}$$

Suppose $\lambda_0 \equiv \lambda_0(G) < 2C \leq 0$. Then

(i) $|P_s \Delta f|_{\ell^2} \leq e^{sC} |\Delta f|_{\ell^2}$

and so

$$\alpha := \lim_{t \to \infty} \frac{1}{2} \int_0^t P_s \Delta f \, ds$$

exists in $\ell_0^2$ giving

$$\lim_{t \to \infty} dP_t f = df + d\alpha$$

in $\ell_1^2$. (The left-hand side is in $\ell_1^2$ by (1).)

(ii) By Lemma 2,

$$\frac{1}{2} \frac{d}{dt} \|(|dP_t f|)^2)\|^2 \leq \langle \left( \frac{1}{2}\Delta - \rho \right) |dP_t f|^2, |dP_t f|^2 \rangle$$

$$\leq \left( \frac{1}{2}\lambda_0 - C \right) \|(|dP_t f|^2)\|^2$$

so that $|dP_t f|_x^2 \to 0$ in $\ell_0^2$ and so in $\ell_0^1$, giving $dP_t f \to 0$ in $\ell_1^2$.
Combining (i) and (ii)

$$df = -d\alpha$$

where $\alpha \in \ell_0^2$. Since $G$ is connected $f$ differs from $\alpha$ by a constant. At least one of $A, B$ must therefore be finite.

Thus $\lambda_0 < 2C$ implies $G$ is 'connected at infinity' and the theorem is proved.  $\square$

**Theorem I'.** *Suppose $g$ is without boundary, i.e., $m(x) > 1$ for all $x$. If $\Delta - 2\rho$ is negative definite on $\ell_0^2$ then $G$ is connected at infinity.*

**Proof.** For such $G$ we have $\rho \leq 0$, and so our hypothesis implies $\lambda_0 < 0$. We can therefore use the same proof as for Theorem I.  $\square$

## §2. Riemannian Manifolds

**A.** Let $M$ be a complete connected Riemannian manifold with dimension $n$. Consider the space $\Omega^1$ of $C^\infty$ 1-forms on $M$, so if $\alpha \in \Omega^1$ it determines a linear map

$$\alpha_x : T_x M \to \mathbb{R}$$

of the tangent space to $M$ at $x$, for each $x \in M$. Using the inner product $\langle \,, \, \rangle_x$ and norm $|\cdot|_x$ on $T_x M$ we obtain

$$|\alpha|: M \to \mathbb{R}$$
$$|\alpha|(x) = |\alpha_x|_x.$$

Let $L^2\Omega^1$ be the space of $L^2$ 1-forms in $\Omega^1$

$$L^2\Omega^1 = \{\alpha \in \Omega^1 : |\alpha| \in L^2(M;\mathbb{R})\},$$

and $\Omega_C^1$ the space of compactly supported forms in $\Omega^1$. The corresponding spaces of functions will be written $\Omega^0$ etc. The operator $d$ maps functions to 1-forms and 1-forms to 2-forms by exterior differentiation and we can form the quotient vector spaces whose additive groups are the first real cohomology groups of the theory indicated:

*de Rham cohomology*

$$H^1(M;\mathbb{R}) = \{\alpha \in \Omega^1 : d\alpha = 0\}/\{df : f \in \Omega^0\}$$

$L^2$ *cohomology*

$$L^2H^1(M;\mathbb{R}) = \{\alpha \in L^2\Omega^1 : d\alpha = 0\}/\{df : f \in L^2\Omega^0\}$$

*cohomology with compact supports*

$$H_C^1(M;\mathbb{R}) = \{\alpha \in \Omega_C^1 : d\alpha = 0\}/\{df : f \in \Omega_C^0\}.$$

To get an idea of the geometric meaning of these spaces, and as a step to the proof of the main result in this section consider the following example: Suppose $K$ is a compact subset of $M$ such that $M - K$ is the disjoint union of two nonempty open sets $A$ and $B$. Let $f: M \to \mathbb{R}$ be $C^\infty$ with $f \mid A \equiv -1$ and $f \mid B \equiv 1$, and denote by $[df]$ its cohomology class in $H_C^1(M;\mathbb{R})$. Since there are natural inclusions $H_C^1(M;\mathbb{R}) \to L^2H^1(M;\mathbb{R}) \to H^1(M;\mathbb{R})$ we can consider $[df]$ in either of the three theories:

**Lemma.** (i) $[df] = 0$ *in* $H^1$.
(ii) $[df] = 0$ *in* $L^2H^1$ *iff at least one of $A$ and $B$ has finite volume.*
(iii) $[df] = 0$ *in* $H_C^1$ *iff at least one of $A$ and $B$ is compact.*

**Proof.** (i) is immediate; for (ii) if $A$ has finite volume set $g(x) = f(x) - 1$ then $g$ vanishes outside the finite volume set $A \cup K$ and so is in

$L^2$. Since $dg = df$ we have $[df] = 0$ in $L^2H^1$. Conversely suppose $df = dg$ for some smooth $g \in L^2(M; \mathbb{R})$. Then $g$ is constant on each component of $M - K$ and differs from $f$ by a constant. This is not possible unless at least one of $A$ and $B$ as finite volume.

The proof of (iii) is essentially the same. $\square$

**B.** There is the Laplace–Beltrami operator

$$\Delta^1 = -(dd^* + d^*d): \Omega^1 \to \Omega^1$$

where $d^*$ is the $L^2$ adjoint of the exterior derivative $d$. For smooth $f$ we see

$$\Delta^1 df = d\Delta f$$

where $\Delta$ is the Laplacian on functions. Forming $\Gamma(f, g)$ and $\Gamma_2(f, g)$ for suitable $f: M \to \mathbb{R}$, just as in §2 we have

$$(f, g)(x) = \langle \nabla f(x), \nabla g(x) \rangle_x$$

(note the lack of the $\frac{1}{2}$ which appeared in the case of graphs) and, [2], p. 188,

$$\Gamma_2(f, g)(x) = \langle \nabla^2 f(x), \nabla^2 g(x) \rangle_x + \mathrm{Ric}_x(\nabla f(x), \nabla g(x)).$$

Here $\mathrm{Ric}_x: T_xM \times T_xM \to \mathbb{R}$ is the Ricci curvature tensor at $x$ and

$$\nabla^2 f(x): T_xM \times T_xM \to \mathbb{R}$$

is the second covariant derivative (or Hessian) of $f$ at $x$. The Ricci tensor is a symmetric bilinear map and for $v \in T_xM$, $\mathrm{Ric}_x(v, v)$ is the sum of the sectional curvatures determined by an orthogonal family of planes in $T_xM$ through $v$.

The formula for $\Gamma_2$ is closely related to the Weitzenbock formula (e.g., see [14]) for $\alpha \in \Omega^1$:

$$\Delta^1 \alpha = \mathrm{trace}\, \nabla^2 \alpha - \mathrm{Ric}(\alpha^{\#}, -)$$

where $\alpha^{\#}$ is the vector field dual to $\alpha$. This is the basis for the use of a 'Feynman–Kac formula' for the solution of the heat equation for 1-forms. Rather than describe the probabilistic solution to the heat equation for forms (given in detail in [19], [12], or [14] for example), let us go directly to the semigroup domination which follows from it. For this set

$$\rho(x) = \inf\big\{\mathrm{Ric}_x(v, v): v \in T_xM, |v|_x = 1\big\}$$

and $C = \inf\{\rho(x): x \in M\}$.

*Assume from now on that $C > -\infty$.*

Let $\{P_t^1: t \geq 0\}$ be the heat semigroup on 1-forms. It can be considered as acting on the Hilbert space $\overline{L^2\Omega^1}$ of $L^2$ one-forms by $P_t^1 = e^{\frac{1}{2}\Delta^1 t}$, (with suitable definition of $\Delta^1$ as a closed operator) or on the space of bounded continuous forms e.g., via the Feynman–Kac formula mentioned above. These definitions agree on the intersection of their domains, e.g., see [13]. Let $\{Q_t: t \geq 0\}$ be the semigroup on functions with generator $\frac{1}{2}\Delta - \frac{1}{2}\rho$ acting on $L^2$ or $L^\infty$. The domination result is: for any bounded continuous 1-form $\alpha$

$$|P_t^1\alpha|_x \leq Q_t(|\alpha|(x))$$
$$\leq e^{-\frac{1}{2}Ct}P_t(|\alpha|)(x)$$

for $\{P_t: t \geq 0\}$ the heat semigroup on functions.

**C.** Let $\lambda_0(M)$ be the upper bound of the spectrum of $\Delta$ on $L^2(M; \mathbb{R})$. The main result of the section is the following, proved with S. Rosenberg [16].

**Theorem II.** [16] *Let $M$ be a complete Riemannian manifold $M$ whose Ricci curvature is bounded below by a constant $C$. Suppose there is a compact subset $K$ of $M$ such that $M - K$ is the disjoint union of two open sets $A, B$ of infinite volume. Then*

$$\lambda_0(M) \geq C.$$

**Proof.** Suppose $\lambda_0(M) < C$. We can assume $\lambda_0(M) < 0$ since $C > 0$ implies that $M$ is compact by a theorem of Myers (e.g., see [24]).

For $A, B, K$ as described take a $C^\infty$ map $f: M \to [0,1]$ with $f(x) = 1$ on $A$ and $f(x) = -1$ on $B$. By Lemma (ii) it suffices to show $df = dg$ for some $L^2$ function $g$ to obtain a contradiction. As for Theorem I,

$$dP_t f = df + \frac{1}{2}d\int_0^t P_s(\Delta f)ds.$$

Note: (i) since $\lambda_0(M) < 0$

$$g := -\lim_{t\to\infty} \frac{1}{2}\int_0^t P_s(\Delta f)ds$$

exists in $L^2(M; \mathbb{R})$ and

(ii) since $\lambda_0(M) < C$

$$|dP_t f| = |P_t^1 df| \le e^{-Ct} P_t |df| \to 0$$

in $L^2(M; \mathbb{R})$.

Now $d$ restricted to the space of $C^1$ functions with compact support has closure which contains $d$ acting on $C^1$ functions $f$ with both $f$ and $df$ in $L^2$, (as observed by Gaffney [17], p. 465). Using this closure, (i) and (ii) imply

$$df = dg.$$

It follows from this that $g$ is smooth; alternatively this follows from the fact that $\Delta g = \Delta f$ (by the definition of $g$).   □

The proof could have been shortened by quoting Donnelly's version of Hodge's theorem [10]. For $\lambda_0(M) < 0$ this tells us that, for $f$ as in the proof

$$df = d\alpha + \theta$$

where $\alpha$, $d\alpha$, $\theta$ are in $L^2$ and $\theta$ is harmonic. The $\lambda_0(M) < C$ assumption will imply $\theta = 0$ by semigroup domination. In fact the proof given essentially incorporated a proof of that version of the Hodge theorem.

By [10] the Hodge theorem also holds if the spectrum of $\Delta^1$ is bounded below zero. We could therefore have expressed our spectral condition in terms of the operator $\Delta - \rho$ and obtained:

**Theorem II'.** *If $\Delta - \rho$ is a negative definite operator on $L^2(M; \mathbb{R})$ then $M$ has at most one infinite volume end.*

This condition is discussed in detail in [16] for the case when $M$ covers a compact manifold.

## §3. Concluding Remarks

**A.** The estimate obtained for $\lambda_0(M)$ should be compared with Cheng's inequality: *If $M$ is complete and $C < 0$ then*

$$\lambda_0(M) \ge \left( \frac{n-1}{4} \right) C.$$

Thus Theorem II is only worthwhile for $n > 5$. On the other hand this does not apply to Theorem II'.

For graphs without boundary and $\overline{m} = \sup m(x) < \infty$, Molchanov showed me the somewhat analogous result that $\lambda_0(G) \ge 2\frac{\sqrt{(\overline{m}-1)}}{\overline{m}} - 1$. (To see this consider $f(x) = \lambda^{r(x)}$ where $r(x)$ is the (obvious) distance of $x$ from some fixed $x_0$ and $0 < \lambda < 1$.) Thus Theorem 1 is not relevant for such $G$.

**B.** Note the difference in the results of I and II: for graphs the inequality obtained was $\lambda_0(G) \geq 2C$ for manifolds it was $\lambda_0(M) \geq C$. This was due to the weaker domination obtained using $\Gamma_2$. It would be interesting to know if it can be improved.

**C.** Presumably it is possible to obtain results for more general weightings on graphs, and possibly for the combinatorial Laplacians used by Dodziuk and Patodi [7], [8]. Work in progress by L. Smits may help here. When the graph $G$ in question is the 1-dimensional skeleton of a triangulation of a complete Riemannian manifold $M$ the hypothesis of Theorem II for $M$ clearly implies the hypothesis of I for $G$.

**Acknowledgements:** This is part of a programme being pursued in collaboration with S. Rosenberg. The results for graphs were worked out together with D. Bakry. Conversations with L. Smits, V. Kaimanovitch and S.A. Molchanov proved very helpful.

## REFERENCES

[1] H. Airault, *Subordination de processus dans le fibré tangent et formes harmoniques*, C.R. Acad. Sci. Paris, Sér. A, **282** (June 14, 1976), 1311–1314.

[2] D. Bakry and M. Emery, *Diffusions hypercontractives*. In *Seminaire de Probabilités XIX, 1983–84*, eds. J. Azéma and M. Yor, Lecture Notes in Math. **1123**, Springer-Verlag, 1985.

[3] D. Bakry and M. Emery, *Propaganda for $\Gamma_2$*. In *From Local Times to Global Geometry, Control and Physics*, ed. K.D. Elworthy, Pitman Research Notes in Math. **150**, Longman, 1986, pp. 39–46.

[4] D. Bakry, *Inégalités de Sobolev faible: un critère $\Gamma_2$*, to appear in Seminaire de Probabilités, L.N.M. Springer, 1990.

[5] P.H. Bérard, *From vanishing theorems to estimating theorems: the Bochner technique revisited*, Bull. Amer. Math. Soc. **19** (1988), 371–406.

[6] A.M. Berthier and B. Gaveau, *Critère de convergence des fonctionnelles de Kac et applications en mécanique quantique et en géométrie*, J. Funct. Anal. **29** (1978), 416–424.

[7] J. Dodziuk, *Finite-difference approach to the Hodge theory of harmonic forms*, Amer. J. Math. **98** (1976), 79–104.

[8] J. Dodziuk and V.K. Patodi, *Riemannian structures and triangulation of manifolds*, J. Indian Math. Soc. **40** (1976), 1–52.

[9] J. Dodziuk and W.S. Kendall, *Combinatorial Laplacians and isoperimetric inequality.* In *From Local Times to Global Geometry, Control and Physics*, ed. K.D. Elworthy, Pitman Research Notes in Math. Series, **150**, Longman, 1986, pp. 68–74.

[10] H. Donnelly, *The differential form spectrum of hyperbolic space*, Manuscripta Math. **33** (1981), 365–385.

[11] H. Donnelly and P. Li, *Lower bounds for the eigenvalues of Riemannian manifolds*, Michigan Math. J. **29** (1982), 149–161.

[12] K.D. Elworthy, *Stochastic differential equations on manifolds*, London Mathematical Society Lecture Notes **70**, Cambridge University Press, Cambridge, 1982.

[13] K.D. Elworthy and S. Rosenberg, *Generalized Bochner theorems and the spectrum of complete manifolds*, Acta Appl. Math. **12** (1988), 1–33.

[14] K.D. Elworthy, *Geometric aspects of diffusions on manifolds.* In *Ecole d'Eté de Probabilités de Saint-Flour XV–XVII, 1985–1987*, ed. P.L. Hennequin, Lecture Notes in Math. **1362**, Springer-Verlag, 1989, pp. 276–425.

[15] K.D. Elworthy and S. Rosenberg, *Spectral bounds and the shape of manifolds near infinity.* In *Proc. IX International Congress on Mathematical Physics, Swansea 1988*, eds. B. Simon, I.M. Davies and A. Truman, Adam Hilger, Bristol & New York, 1989, pp. 369–373.

[16] K.D. Elworthy and S. Rosenberg, *Manifolds with wells of negative curvature*, Invent. Math., **103** (1991), 471–495.

[17] M.P. Gaffney, *The heat equation method of Milgram and Rosenbloom for open Riemannian manifolds*, Annals of Mathematics **60**, No. 3 (1954), 458–466.

[18] P. Gerl, *Random walks on graphs with a strong isoperimetric property*, Jour. Theor. Prob. **1**, No. 2 (1988), 171–187.

[19] N. Ikeda and S. Watanabe, *Stochastic Differential Equations and Diffusion Processes*, Kodansha, Tokyo: North-Holland, Amsterdam, New York, Oxford, 1981.

[20] K. Itô, *The Brownian motion and tensor fields on a Riemannian manifold*, Proc. Internat. Congr. Math. (Stockholm, 1962), Inst. Mittag-Leffler, 1963, pp. 536–539.

[21] P. Malliavin, *Formule de la moyenne pour les formes harmoniques*, J. Funct. Anal. **17** (1974), 274–291.

[22] P. Malliavin, *Annulation de cohomologie et calcul des perturbations dans $L^2$*, Bull. Sc. Math. **100** (1976), 331–336.

[23] A. Meritet, *Théorème d'annulation pour la cohomologie absolue d'une variété Riemannienne abord*, Bull. Sc. Math. **103** (1979), 379–400.

[24] J. Milnor, *Morse Theory*, Annals of Math. Studies **51**, Princeton University Press, Princeton, 1963.

[25] J. Vauthier, *Théorèmes d'annulation et de finitude d'espace de 1-formes harmoniques sur une variété de Riemann ouverte*, Bull. Sc. Math. **103** (1979), 129–177.

Mathematics Institute
University of Warwick
Coventry CV4 7AL, ENGLAND

# The Thom class of Mathai and Quillen and probability theory

EZRA GETZLER

Department of Mathematics,
MIT, Mass. 02139

As we will explain in this talk, the construction by Mathai and Quillen of explicit differential form representatives of the Thom class and Euler class of a vector bundle [5] gives a framework for understanding in unified way a number of ideas in stochastic differential geometry. We will also show briefly how Witten's topological quantum field theories fit into this formalism. For the moment, the picture that he envisages is inaccessible to rigourous methods; but even in the more humdrum world of Brownian motion, the point of view presented here illuminates quite a few of the other talks that were given at this conference.

For a more thorough presentation of the mathematical portion of this talk, we refer the reader to Chapters 1 and 7 of the book [2].

## 1. THE THOM CLASS

Recall that the de Rham complex of differential forms on a manifold $M$ is consists of the infinite dimensional vector spaces $\mathcal{A}^i(M)$ of $i$-forms, with the exterior differential

$$d : \mathcal{A}^i(M) \to \mathcal{A}^{i+1}(M).$$

Let $E \xrightarrow{\pi} M$ be an orientable vector bundle with fibre $\mathbb{R}^m$ over a manifold $M$. A **Thom class** for the bundle $E$ is a differential form $\mu(E) \in \mathcal{A}^m(E)$ on $E$ such that

(1) $\mu(E)$ is closed;
(2) $\pi_*\mu(E)$, the integral of $\mu(E)$ over the fibres of $\pi$, equals 1.

To construct a Thom class, we suppose that we are given three pieces of data:

(1) an orientation of $E$, that is, a nowhere-vanishing section of the highest exterior power $\Lambda^m E$ of $E$;
(2) a metric on the bundle $E$, that is, a bundle map

$$E \otimes E \to M \times \mathbb{R}$$

which induces positive definite inner products $(\cdot, \cdot)_x$ on each fibre $E_x$ of $E$;
(3) a connection $D$ compatible with the metric $(\cdot, \cdot)$, that is, a map

$$D : \mathcal{A}^i(M, E) \to \mathcal{A}^{i+1}(M, E)$$

such that

$$D(\alpha \wedge \beta) = (d\alpha) \wedge \beta + (-1)^j \alpha \wedge (D\beta)$$

for all $\alpha \in \mathcal{A}^i(M)$ and $\beta \in \mathcal{A}^j(M, E)$ and such that

$$d(s_1, s_2) = (Ds_1, s_2) + (s_1, Ds_2)$$

for all $s_1$ and $s_2 \in \Gamma(M, E)$.

Since $E$ has a metric, we can reformulate the orientation of the bundle $E$ as an isometry

$$\mathsf{B}\{\cdot\} : \Lambda^m E \to M \times \mathbb{R}$$

of the highest exterior power of $E$ with the trivial line bundle over $M$. The map $\mathsf{B}\{\cdot\}$ is the **Berezin integral**.

Let $\pi^* E$ be the pull-back of the bundle $E$ over $M$ to a bundle over $E$, whose fibre at $e \in E$ is $E_{\pi(e)}$; this bundle has a metric $\pi^*(\cdot,\cdot)$ with compatible connection $\pi^* D$; we will usually write these as $(\cdot,\cdot)$ and $D$. Let $\Lambda^* \pi^* E \to E$ be the bundle of exterior algebras of $\pi^* E$, and let $\mathcal{A}^*(E, \Lambda^* \pi^* E)$ be the algebra of differential forms on $E$ with values in $\Lambda^* \pi^* E$; it is a bigraded algebra, with graded subspaces

$$\mathcal{A}^{i,j} = \mathcal{A}^i(E, \Lambda^j \pi^* E).$$

The Berezin integral $\mathsf{B}\{\cdot\} : \Lambda^m E \to M \times \mathbb{R}$ extends to a linear form

$$\mathsf{B}\{\cdot\} : \mathcal{A}^i(E, \Lambda^j \pi^* E) \to \mathcal{A}^i(E)$$

which vanishes unless $j = m$.

Let $x$ be the tautological section of $\pi^* E$, that is, the section which maps a point $e \in E$ to the corresponding point $e \in \pi^* E_e = E_{\pi(e)}$. We may think of $x$ as an element of $\mathcal{A}^{0,1}$, and its covariant derivative $Dx$ is an element of $\mathcal{A}^{1,1}$. Define an operator $\iota(x) : \mathcal{A}^{i,j} \to \mathcal{A}^{i,j-1}$, characterized by the following properties:

(1) if $w \in \mathcal{A}^{0,1} = \Gamma(E, \pi^* E)$, then $\iota(x)w = (x, w)$ (for example, $\iota(x)x = |x|^2$);
(2) $\iota(x)$ is a derivation, that is,

$$\iota(x)(\alpha \wedge \beta) = (\iota(x)\alpha) \wedge \beta + (-1)^{i+j}\alpha \wedge (\iota(x)\beta)$$

for $\alpha \in \mathcal{A}^{i,j}$ and $\beta \in \mathcal{A}^{k,l}$.

Let $D_t$ be the operator $D + t\iota(x)$, where $t > 0$.

LEMMA 1.1. *If $\alpha \in \mathcal{A}^*(E, \Lambda^* \pi^* E)$, then $d\mathsf{B}\{\alpha\} = \mathsf{B}\{D_t\alpha\}$.*

PROOF: This follows from the fact that $D$ is compatible with the metric, so that $d\mathsf{B}\{\alpha\} = \mathsf{B}\{D\alpha\}$, combined with the obvious fact that $\iota(x)\alpha$ has no component in $\mathcal{A}^*(E, \Lambda^m \pi^* E)$. ∎

The curvature $\Omega = D^2$ of the connection $D$ on $E$ is an element of $\mathcal{A}^2(M, \mathfrak{so}(E))$, where $\mathfrak{so}(E)$ is the bundle of skew-symmetric endomophisms of $E$. We may identify the bundle $\mathfrak{so}(E)$ with $\Lambda^2 E$; if $A \in \mathfrak{so}(E_y)$ and $e_i$ is an orthonormal basis of $E_y$, we map $A$ to

$$\sum_{1 \leq i < j \leq m} (e_i, Ae_j)\, e_i \wedge e_j \in \Lambda^2 E_y,$$

so that $\iota(x)A = -Ax$. The pull-back $\pi^* \Omega$ of $\Omega$ from $M$ to $E$ is then an element of $\mathcal{A}^{2,2}$, which we will usually denote by $\Omega$. Consider the following differential form (here, $t$ is a positive real number):

$$\omega_t = \tfrac{1}{2}t^2|x|^2 + tDx + \Omega.$$

LEMMA 1.2. *The following formula holds:* $D_t\omega_t = 0$.

PROOF: Note that $D\Omega = 0$, by Bianchi's identity, and that $D|x|^2 = -2\iota(x)Dx$. From this, we see that

$$D\omega_t = \frac{t^2}{2}D|x|^2 + tD^2x = -\iota(x)\left(t^2Dx + t\Omega\right) \in \mathcal{A}^{1,0} \oplus \mathcal{A}^{2,1}.$$

The lemma follows, since it is clear that $\iota(x)|x|^2 = 0$. ∎

Suppose $f$ is a polynomial in one variable. It is easy to see that $f(\omega_t)$ is the element of $\mathcal{A}^{*,*}$ given by the Taylor expansion

$$f(\omega_t) = \sum_{k=0}^{m} \frac{f^{(k)}(t^2|x|^2/2)}{k!}(tDx + \Omega)^k.$$

(The expansion terminates at $k = m$ because $(tDx + \Omega)^k = 0$ for $k > m$.) We adopt this formula as the definition of $f(\omega_t)$ for any smooth function on $\mathbb{R}$.

PROPOSITION 1.3.

(1) *The differential form* $\mathsf{B}\{f(\omega_t)\}$ *is a closed $m$-form on* $E$.
(2) *If $f$ decays at infinity, the integral over the fibres* $\pi_* f(\omega_t)$ *is the constant*

$$\pi_*\mathsf{B}\{f(\omega_t)\} = (-1)^{m(m+1)/2}\int_{\mathbb{R}^m} f^{(m)}(|x|^2/2)\,dx.$$

(3) *(transgression formula)*

$$\frac{d\mathsf{B}\{f(\omega_t)\}}{dt} = d\mathsf{B}\{xf'(\omega_t)\}$$

PROOF: To see that $\mathsf{B}\{f(\omega_t)\} \in \mathcal{A}^m(E)$, we simply note that

$$f(\omega_t) \in \sum_{k=0}^{m} \mathcal{A}^{k,k},$$

and hence that only the component in $\mathcal{A}^{m,m}$ contributes to the Berezin integral.

The proof that $\mathsf{B}\{f(\omega_t)\}$ is closed is an easy consequence of Leibniz's rule and Lemmas 1.1 and 1.2:

$$d\mathsf{B}\{f(\omega_t)\} = \mathsf{B}\{f'(\omega_t)D_t\omega_t\} = 0.$$

The formula for $\pi_*\mathsf{B}\{f(\omega_t)\}$ may be checked in the special case in which $M$ is a point, and hence we may take $E$ to equal $\mathbb{R}^m$. If $x_i$ is the standard orthonormal basis of $\mathbb{R}^m$, we see that

$$Dx = \sum_{i=1}^{m} dx_i \otimes x_i,$$

and hence that

$$(Dx)^m = m!(dx_1 \otimes x_1)\dots(dx_m \otimes x_m)$$
$$= (-1)^{m(m+1)/2}m!(x_1 \wedge \dots \wedge x_m) \otimes (dx_1 \dots dx_m).$$

We see that

$$\mathsf{B}\{f(\omega_t)\} = t^m\mathsf{B}\left\{f^{(m)}(t^2|x|^2/2)(dx_1 \otimes x_1)\dots(dx_m \otimes x_m)\right\}$$
$$= (-1)^{m(m+1)/2}t^m\,\mathsf{B}\left\{f^{(m)}(t^2|x|^2/2)x_1\dots x_m\right\}dx_1\dots dx_m$$
$$= (-1)^{m(m+1)/2}t^m\,f^{(m)}(t^2|x|^2/2)\,dx_1\dots dx_m,$$

since all of the other terms in the Taylor expansion have vanishing Berezin integral. From this, (2) follows easily.

The proof of the transgression formula is similar to the proof that $d\mathsf{B}\{f(\omega_t)\} = 0$. On the one hand,

$$\frac{d\mathsf{B}\{f(\omega_t)\}}{dt} = \mathsf{B}\{(t|x|^2 + Dx)f(\omega_t)\},$$

while on the other,

$$d\mathsf{B}\{xf'(\omega_t)\} = \mathsf{B}\{(D_tx)f(\omega_t)\} = \mathsf{B}\{(Dx + t|x|^2)f(\omega_t)\}. \quad \blacksquare$$

Being probabilists, we are especially fond of the Gaussian function, so we choose as our Thom class the following differential form:

$$\mu_t(E) = (-1)^{m(m-1)/2}(2\pi)^{-m/2}\mathsf{B}\{e^{-\omega_t}\} \in \mathcal{A}^m(E).$$

Let $s$ be a section of the bundle $E$. We can form the pull-back $s^*\mu_t(E) \in \mathcal{A}^m(M)$; it is called the **Euler class** of the bundle $E$. The following formula for the differential form $s^*\mu_t(E)$ is obvious from the definition of $\mu_t(E)$:

$$s^*\mu_t(E) = (-1)^{m(m-1)/2}(2\pi)^{-m/2}\mathsf{B}\left\{e^{-t^2|s|^2/2-tDs-\Omega}\right\},$$

where $t^2|s|^2/2 + tDs + \Omega/2$ is thought of as a section of $\mathcal{A}^*(M, \Lambda^*E)$, and $\mathsf{B}\{\cdot\}$ is the Berezin integral from $\mathcal{A}^*(M, \Lambda^*E)$ to $\mathcal{A}^*(M)$.

PROPOSITION. *The cohomology class of the differential form $s^*\mu_t(E) \in \mathcal{A}^m(M)$ is independent of the section $s$.*

PROOF: Let $s_t = s + \tau$ be an affine one-parameter family of sections of $E$. We see that

$$(-1)^{m(m-1)/2}(2\pi)^{m/2}\frac{ds_\tau^*\mu_t(E)}{d\tau} = \frac{d}{d\tau}\mathsf{B}\left\{e^{-t^2|s_\tau|^2/2-tDs_\tau-\Omega}\right\}$$
$$= -s_\tau^*\mathsf{B}\left\{(t^2(x,u) + tDu)e^{-t^2|x|^2/2-tDx-\Omega}\right\}$$
$$= -s_\tau^*\mathsf{B}\left\{(tD_tu)e^{-t^2|x|^2/2-tDx-\Omega}\right\}$$
$$= -ts_\tau^*d\mathsf{B}\left\{ue^{-t^2|x|^2/2-tDx-\Omega}\right\}$$
$$= -td\mathsf{B}\left\{ue^{-t^2|s_\tau|^2/2-tDs_\tau-\Omega}\right\} \quad \blacksquare$$

## 2. THE GAUSS-BONNET THEOREM

In this section, we explain the geometric significance of the Euler class: its Poincaré dual represents the homology class of the zero-set of a non-degenerate section of $E$. We will assume for simplicity that $M$ is compact, since otherwise we would have to formulate general conditions on the section $s$ to increase at infinity, and these are better done on a case-by-case basis.

Suppose that the section $s \in \Gamma(M, E)$ has the special property that its zero-set $M_0$ is a submanifold of $M$, and furthermore that $\nabla s \in \Gamma(M, \mathrm{Hom}(TM, E))$ is surjective along $M_0$; we call such a section **non-degenerate**. In particular, it follows that $\dim(M_0) = \dim(M) - m$. Let $\delta_{M_0}$ be the current on $M$ whose value on a differential form $\alpha \in \mathcal{A}^*(M)$ is

$$\langle \delta_{M_0}, \alpha \rangle = \int_{M_0} \alpha \big|_{M_0}.$$

Thus, $\delta_{M_0}$ vanishes on $\alpha$ unless $\alpha \in \mathcal{A}^{\dim(M)-m}(M)$. The following theorem explains the significance of the Euler class.

THEOREM 2.1. *The differential forms $s^* \mu_t(E) \in \mathcal{A}^m(M)$ converge, as $t \to \infty$, to a current of the form*

$$\lim_{t \to \infty} s^* \mu_t(E) = \varepsilon \delta_{M_0},$$

*where $\varepsilon : M_0 \to \{\pm 1\}$ is a continuous function which measures whether the map $\nabla s$ preserves or reverses orientation.*

PROOF: By (1.3), we see that

$$s^* \mu_t(E) = (-1)^{m(m-1)/2} (2\pi)^{-m/2} \mathsf{B} \left\{ e^{-t^2 |s|^2 / 2 - tDs - \Omega} \right\}.$$

To take the limit as $t \to \infty$ in this formula, we observe that there is clearly no contribution from the region in which $|s| >$, for some small $c > 0$. Thus, the limiting current is supported on the submanifold $M_0$. We may assume that $M$ is an open subset of $\mathbb{R}^n$, parametrized by coordinates

$$(x_1, \ldots, x_m, y_1, \ldots, y_{n-m}),$$

that $M_0 = U \cap \mathbb{R}^m$, where

$$\mathbb{R}^m = \{(x, y) \in \mathbb{R}^n \mid x = 0\}$$

and that $E$ is the trivial bundle with fibre $\mathbb{R}^m$. In a neighbourhood of $M_0$, we see that the endomorphism $\nabla s : M \to \mathrm{Hom}(\mathbb{R}^n, \mathbb{R}^m)$ is surjective, and hence, possibly replacing $M$ by a smaller neighbourhood of $M_0$ in $\mathbb{R}^n$, we may choose as a frame of the bundle $E$ the sections

$$e_i = D_{\partial/\partial x_i} s, \quad 1 \le i \le m.$$

For the moment, assume that the resulting frame of $E$ is oriented.

Let $A_{ij} = (e_i, e_j)$ be the $m \times m$-matrix of inner products in this frame. By changing the coordinate system on $M$ to

$$\tilde{x}_i = \left( A(0, y)^{-1/2} \right)_{ij} x_j,$$

we may assume that $(e_i, e_j) = \delta_{ij} + O(|x|)$.

We see that the section $s$ may be written

$$s(x, y) = \sum_{i,j=1}^{m} x_i f_{ij}(x, y) e_j(x, y),$$

where $f_{ij} \in C^\infty(M)$ are functions satisfying $f_{ij}(0, y) = \delta_{ij}$. It follows that

$$Ds = \sum_{j=1}^{m} \left( \sum_{i=1}^{m} g_{ij} dx_i \otimes e_j + \sum_{i=1}^{n-m} h_{ij} dy_i \otimes e_j \right),$$

where $g_{ij}(0, y) = \delta_{ij}$ and $h_{ij}(0, y) = 0$. Finally, let $\Omega_0 \in \mathcal{A}^2(M_0, \Lambda^2 E)$ be the restriction of the curvature $\Omega \in \mathcal{A}^2(M, \Lambda^2 E)$ to $M_0 \subset M$.

If $\varphi \in C_0^\infty(M)$ is a function on $M$ with compact support, we have

$$\int_M \varphi \, \mu_t \, dx \, dy = (-1)^{m(m-1)/2} (2\pi)^{-m/2} \int_M \varphi \, \mathrm{B} \left\{ e^{-t^2 |s|^2/2 - tDs - \Omega} \right\} dx \, dy.$$

Pull this differential form back by the map $\rho_t(x, y) = (t^{-1} x, y)$. Since the integral of a differential form is invariant under pull-back, we see that

$$\int_M \varphi \, \mu_t \, dx = (-1)^{m(m-1)/2} (2\pi)^{-m/2}$$
$$\int_{\mathbb{R}^m \times M_0} \varphi(t^{-1} x, y) \, \mathrm{B} \left\{ e^{-t^2 |s(t^{-1} x, y)|^2/2 - t\rho_t^* Ds - \rho_t^* \Omega} \right\} dx \, dy.$$

Under these deformations, the exponent becomes

$$\frac{1}{2} \sum_{i=1}^{m} x_i^2 + \sum_{i=1}^{m} dx_i \otimes e_i + \Omega_0(0, y) + O(t).$$

By the dominated convergence theorem, the integral over $\mathbb{R}^m \times M_0$ converges to

$$(2\pi)^{-m/2} \int_{\mathbb{R}^m \times M_0} \varphi(0, y) \, e^{-|x|^2/2} \mathrm{B} \left\{ \exp\left( -\sum_{i=1}^{m} dx_i \otimes e_i(0, y) - \Omega_0(0, y) \right) \right\} dx \, dy.$$

It is easily seen that $\Omega_0$ cannot contribute to this integral, since any term involving $\Omega_0$ will not have enough powers of $dx_i$ to give a non-zero answer. This shows that this limit equals $\int_{M_0} \varphi(0, y) \, dy$ as desired.

This handles the case in which the basis $e_i$ is an oriented basis of $E$. If it is not an oriented basis, there is an additional sign $\varepsilon = -1$, but otherwise the above calculation is unchanged. ∎

Theorem 2.1 implies as a special case the Gauss-Bonnet-Chern theorem. Let $M$ be a compact, oriented, even-dimensional Riemannian manifold, and let $E = T^*M$. We choose a Morse function $f$ on $M$, and consider the section $s = df$. By Proposition 1.4, the differential forms $s^*\mu_t(E)$ are cohomologous. Setting $t = 0$, we obtain the differential form

$$s^*\mu_0(T^*M) = (-2\pi)^{-m/2} \mathsf{B}\left\{e^{-\Omega}\right\} = (2\pi)^{-m/2} \operatorname{Pf}(\Omega),$$

where $\operatorname{Pf}(\Omega) = \mathsf{B}\left\{e^\Omega\right\}$ is the Pfaffian of $\Omega$. On the other hand, as $t \to \infty$, the differential forms $s^*\mu_t(T^*M)$ converge in the distributional sense to the differential form

$$s^*\mu_\infty(T^*M) = \sum_{df(x)=0} \operatorname{sgn}\det(\operatorname{Hess}_x(f))\,\delta_x.$$

Here, $\operatorname{Hess}_x(f) = \nabla df(x)$ is the Hessian of the function $f$ at $x \in M$, which may be considered to be an element of $\operatorname{End}(T_x^*M)$.

In this way, we see that

$$\int_M s_0^*\mu(T^*M) = (2\pi)^{-m/2} \int_M \operatorname{Pf}(\Omega)$$

and

$$\lim_{t\to\infty} \int_M s^*\mu_t(T^*M) = \sum_{df(x)=0} \operatorname{sgn}\det(\operatorname{Hess}_x(f))$$

are equal. By Morse Theory, this second sum is just the Euler number of the manifold, and we obtain the Gauss-Bonnet-Chern theorem.

### 3. An Infinite-Dimensional Example

In this section, we will see that the Euler class constructed in Section 1 can be used to understand the functional integral of the Laplace-Beltrami operator on a compact manifold $X$. This is a simple example of a topological quantum field theory: the Hilbert space is the space of harmonic forms on $X$, which is a topological invariant of the manifold.

If $X$ is a compact orientable Riemannian manifold, let $LX$ be the loop space of $X$; this is the Hilbert manifold of all maps $\gamma : S^1 \to X$ of finite energy,

$$E(\gamma) = \int_{S^1} |\dot\gamma(t)|^2\, dt < \infty.$$

The tangent space $T_\gamma LX$ of the loop space at a loop $\gamma$ is the space of all finite energy tangent vector fields along $\gamma$.

If $SO(X)$ is the orthonormal frame bundle of $X$, then the loop space $LSO(X)$ is a principal bundle over $LX$ with structure group $LSO(n)$. The tangent bundle of $LX$ is the bundle associated to $LSO(X)$ with fibre $L_1^2(S^1, \mathbb{R}^n)$; however, the action of

$L\,\mathrm{SO}(n)$ is not unitary, so this representation does not give the Hilbert bundle structure of $T(LX)$.

The principal bundle $L\,\mathrm{SO}(X)$ has a connection which is derived from the Levi-Civita connection on $\mathrm{SO}(X)$ as follows: if $\theta \in \mathcal{A}^1(\mathrm{SO}(X); \mathfrak{so}(n))$ is the connection one-form on $\mathrm{SO}(X)$, then the connection form on $L\,\mathrm{SO}(X)$ is just the element of $\mathcal{A}^1(L\,\mathrm{SO}(X); L\mathfrak{so}(n))$ given by integrating $\theta$ around the circle. It is not very difficult to calculate the curvature of this connection; it is just the element of $\mathcal{A}^2(L\,\mathrm{SO}(X); \mathfrak{so}(n))$ given by integrating the curvature form $R \in \mathcal{A}^2(\mathrm{SO}(X); \mathfrak{so}(n))$ around the circle:

$$(\Omega(X,Y)Z, W) = \int_{S^1} \big( R_{\gamma(t)}(X_t, Y_t)Z_t, W_t \big)\, dt.$$

Over the manifold $LX$, we will also consider the Hilbert bundle $E$ whose fibre at $\gamma \in LX$ is the space of square-integrable vector fields along $\gamma$; this is the bundle associated to $L\,\mathrm{SO}(X)$ with fibre $L^2(S^1, \mathbb{R}^n)$. On this Hilbert space, the structure group $L\,\mathrm{SO}(n)$ acts in a unitary fashion, so that $E$ is a Hilbert bundle with compatible covariant derivative, derived from the connection on $L\,\mathrm{SO}(X)$.

There is a smooth section of the bundle $E$, given by taking a finite energy loop $\gamma$ to the $L^2$-vector field $\dot{\gamma} \in E_\gamma$, the tangent vector field to the loop. It is clear that the zero-set of this section is the space of constant loops, which we may think of as a manifold $X \subset LX$. Another class of sections of $E$ are determined by smooth functions $f \in C^\infty(M)$: we map $\gamma \in LX$ to the tangent vector field

$$(t \mapsto \operatorname{grad} f(\gamma(t))) \in E_\gamma.$$

Call this section $\operatorname{grad} f$.

LEMMA 3.1. *If $f$ is a Morse function, the section $s_f = \dot{\gamma} + \operatorname{grad} f$ has as its zero set the finite set of constant loops taking values at a critical point of $f$.*

PROOF: Let us calculate the $L^2$-norm of $\dot{\gamma} + \operatorname{grad} f$:

$$\int_{S^1} |\dot{\gamma}(t) + \operatorname{grad}_{\gamma(t)} f|^2\, dt = \int_{S^1} \Big( |\dot{\gamma}(t)|^2 + 2(\dot{\gamma}(t), \operatorname{grad}_{\gamma(t)} f) + |\operatorname{grad}_{\gamma(t)} f|^2 \Big)\, dt$$

$$= E(\gamma) + \int_{S^1} V(\gamma(t))\, dt,$$

where $V = |\operatorname{grad} f|^2 \in C^\infty(X)$. Here we have used the fact that the cross-term vanishes by integration by parts:

$$\int_{S^1} (\dot{\gamma}(t), \operatorname{grad}_{\gamma(t)} f)\, dt = \int_{S^1} \frac{d}{dt} f(\gamma(t))\, dt = 0. \quad \blacksquare$$

This calculation is familiar from the study of the Nicolai map (see [3] for references, and also [4]). Indeed, we see that the Nicolai map is just a section of a vector bundle over the space of fields, and that the supersymmetric quantum field theories with Nicolai maps are just those whose finite temperature (periodic time) functional integral is an Euler class: these are the **topological quantum field theories**.

Let us now try to make some sense of the Euler class of $E$ obtained by pulling back the Thom form by the section $s_f$. According to the formulas of Section 1, this is the differential form on $E$ given by the formula

$$(-1)^{m(m-1)/2}(2\pi)^{-m/2}\exp\left(-\tfrac{1}{2}\int_{S^1}V(\gamma(t))\right)e^{-E(\gamma)/2}\mathrm{B}\left\{e^{-Ds_f-\Omega}\right\}.$$

The difficulty with this expression is that since the bundle $E$ is infinite dimensional, $\mathrm{B}\{\cdot\}$ makes no sense. The constant $(-1)^{m(m-1)/2}(2\pi)^{-m/2}$ is also meaningless, of course; we are not even sure why the rank of the bundle $E$ is even. Of course, the solution of these problems is tied together.

Let us look at the case in which $X = \mathbb{R}^n$, which was discussed in [3]. Here, $\Omega = 0$ while $Ds_f$ may be thought of as the infinite-dimensional generalization of

$$\sum_{ij}A_{ij}dx_i\otimes e_j,$$

where $A_{ij}$ is the operator

$$a\in C^\infty(S^1,\mathbb{R}^n)\mapsto\frac{da(t)}{dt}+\mathrm{Hess}_{\gamma(t)}(f)a(t).$$

If we set $f = 0$, we may interpret the Euler class as being the Brownian bridge measure on $L\mathbb{R}^n$, thought of as a volume form. For this measure to be defined, we must of course replace the manifold $L\mathbb{R}^n$ by the space of all continuous loops in $\mathbb{R}^n$. However, we will be somewhat lax in keeping track of this.

The discussion of [3] justifies the following result (which is more a definition than a theorem, since it relates two objects, one of which is a ill-defined). Let $d_f = e^{-f}\cdot d\cdot e^f$, and let $\Delta_f$ be Witten's twisted Laplace-Beltrami operator

$$\Delta_f = (d_f + d_f^*)^2 = \Delta + |\operatorname{grad} f|^2 + \mathrm{Hess}(f),$$

acting on $\mathcal{A}^*(\mathbb{R}^n)$, where $\mathrm{Hess}(f)$ is the derivation of the algebra of differential forms which vanishes on functions and equals $\mathrm{Hess}(f)\alpha$ for a one-form $\alpha$ (see [6]).

If $\alpha = f_0 df_1\ldots df_k$ is a differential form on $\mathbb{R}^n$, let $c(\alpha)$ be the operator of Clifford multiplication on differential forms given by the formula

$$c(f_0 df_1\ldots df_k) = \frac{1}{k!}\sum_{\sigma\in S_k}(-1)^{\varepsilon(\sigma)}f_0\left(\varepsilon(df_{\sigma_1})-\varepsilon(df_{\sigma_1})^*\right)\ldots\left(\varepsilon(df_{\sigma_k})-\varepsilon(df_{\sigma_k})^*\right).$$

Finally, let Str be the supertrace on the Hilbert space of $L^2$-differential forms on $\mathbb{R}^n$, that is,

$$\mathrm{Str}_{\mathcal{A}(\mathbb{R}^n)}(A) = \mathrm{Tr}_{\mathcal{A}^{2\bullet}(\mathbb{R}^n)}(A) - \mathrm{Tr}_{\mathcal{A}^{2\bullet+1}(\mathbb{R}^n)}(A).$$

A cylinder form is a differential form on $L\mathbb{R}^n$ obtained by taking differential forms $\{\alpha_i\in\mathcal{A}^*(\mathbb{R}^n)\mid 1\le i\le n\}$, and times $0\le t_1<\cdots<t_n<1$, and forming

$$\gamma_{t_1}^*\alpha_1\ldots\gamma_{t_n}^*\alpha_n\in\mathcal{A}(L\mathbb{R}^n).$$

"THEOREM". *The integral of a cylinder form against the Euler form $s_f^* \mu(E)$ is given by the formula*

$$\int_{L\mathbb{R}^m} \gamma_{t_1}^* \alpha_1 \ldots \gamma_{t_n}^* \alpha_n \wedge s_f^* \mu(E)$$

$$= \mathrm{Str}_{A(\mathbb{R}^m)}\big(e^{-t_1 \Delta_f} c(\alpha_1) e^{-(t_1-t_2)\Delta_f} \ldots e^{-(t_{n-1}-t_n)\Delta_f} c(\alpha_n) e^{-(1-t_n)\Delta_f}\big),$$

*where $e^{-t\Delta}$ is the heat kernel of the operator $\Delta_f$.*

Emboldened by this success, it is clear that there is only one plausible possibility for the Euler class on $LX$ for arbitrary $X$, given by exactly the same formula as above except that we replace $\mathbb{R}^n$ throughout by the Riemannian manifold $X$. We now see that the curvature $\Omega$ in the exponential used to define the Thom class corresponds in a precise way to the curvature in the Feynman-Kac formula for the Laplace-Beltrami operator, which may be seen by writing out the Weitzenböck formula for $\Delta_f$:

$$\Delta_f = \nabla^* \nabla + |\mathrm{grad}\, f|^2 + \sum_{ijkl} R_{ijkl}\, \varepsilon^i \iota^j \varepsilon^k \iota^l + \sum_{ij} \partial_i \partial_j f\, \varepsilon^i \iota^j.$$

(This formula is with respect to an orthonormal frame of the cotangent bundle.) In this way, we can understand better some of the calculations made by Alvarez-Gaumé in his heuristic proof of the Gauss-Bonnet-Chern theorem for $X$ [1]: he was repeating on the loop-space $LX$ the calculations of the last section.

## 4. ANOTHER EXAMPLE: DONALDSON POLYNOMIALS

We will briefly indicate one more example of the formalism of Euler classes, introduced by Witten to give an analytic approach to Donaldson polynomials [7]. Let us recall the setting of the Yang-Mills equation in four dimensions. We will ignore questions of Sobolev spaces and regularity: this section is physics!

Let $X$ be a compact oriented four dimensional Riemannian manifold, and let $P \to X$ be a principal $G$-bundle over $X$, where $G$ is a compact Lie group, with Lie algebra $\mathfrak{g}$. Let $\mathcal{A}$ be the space of all connections on the bundle $P$: if we choose a background connection $\nabla$, this space may be realized as the affine space

$$\nabla + \mathcal{A}^1(X, P \times_{\mathrm{ad}} \mathfrak{g}).$$

Let $\mathcal{G}_*$ be the restricted gauge group of the bundle $P$: this is the space of sections

$$\mathcal{G}_* = \{g \in \Gamma(X, P \times_{\mathrm{Ad}} G) \mid g(*) = 1\},$$

where $*$ is a basepoint in $X$. It acts freely on the space $\mathcal{A}$ by the action

$$g \cdot A = g^{-1} A g + g^{-1} dg.$$

We will think of $\mathcal{A}$ as a principal bundle, with structure group $\mathcal{G}_*$ and base $\mathcal{B}$:

$$\mathcal{G}_* \longrightarrow \mathcal{A}$$
$$\downarrow$$
$$\mathcal{B}$$

Incidentally, this shows that $B$ is a classifying space for the group $\mathcal{G}_*$.

The second exterior power $\Lambda^2 T^* X$ of the cotangent bundle of $X$ splits into two pieces

$$\Lambda^2 T^* X \cong \Lambda^+ T^* X \oplus \Lambda^- T^* X,$$

the self-dual and anti-self-dual bundles respectively: these are defined by the equation

$$\star \alpha = \pm \alpha,$$

where $\star$ is the Hodge dual operator. Let $\mathcal{H} = \Omega^-(X, P \times_{\mathrm{ad}} \mathfrak{g})$ be the space of anti-self-dual two-forms with values in the adjoint bundle. Since the gauge group acts on the bundle $P \times_{\mathrm{ad}} \mathfrak{g}$, we may form an associated bundle

$$E = \mathcal{A} \times_{\mathcal{G}_*} \mathcal{H} \to \mathcal{B}.$$

This bundle has a natural metric, since $\mathcal{G}_*$ preserves the $L^2$-metric on $\mathcal{H}$, and a natural connection compatible with this metric, coming from the natural invariant Riemannian structure on $\mathcal{A}$. The curvature of this connection is not pretty: it may be given in terms of a Green's kernel associated to the covariant derivative $\nabla + A$ acting on the space $\Omega^1(X, P \times_{\mathrm{ad}} \mathfrak{g})$.

To construct an Euler class, we need a section of the bundle $E$. This is easily obtained. The curvature $F$ of the connection $\nabla + A$ is an element of $\Omega^2(X, P \times_{\mathrm{ad}} \mathfrak{g})$, so its anti-self-dual component $F^-$ lies in $\Omega^-(X, P \times_{\mathrm{ad}} \mathfrak{g})$. Since the curvature transforms correctly under the gauge group, we see that $F^-$ defines a section of the bundle $E$, whose zero-set is the space of self-dual connections. In [7], Witten presents evidence that the Donaldson polynomials, which are invariants of the bundle $P \to X$, may be obtained by applying a formal version of Theorem 2.1 to this situation. It would be interesting to understand the extent to which his arguments are more than formal.

For the sake of accuracy, we should mention one further point: the above structures are all acted on by a finite-dimensional group, namely the quotient group $\mathcal{G}/\mathcal{G}_* \cong G$, where $\mathcal{G}$ is the gauge group with no restriction on its value at the basepoint $*$. The spaces $\mathcal{A}$, $\mathcal{H}$, $\mathcal{B}$ and $E$ all carry compatible actions of this group, and the section $F^-$ of $E$ is equivariant; the quotient of its zero-set by the action of $G$ is the moduli space $\mathcal{M}$. However, the action of $G$ on $\mathcal{B}$ is not in general free. It turns out that to formulate correctly the Donaldson invariants, one must consider a generalization of the formalism of Section 1, in which we replace the algebra of differential forms on $E$ by the algebra of equivariant differential forms: this is the space of maps from the Lie algebra $\mathfrak{g}$ of $G$ to $\Omega(E)$ invariant under the action of $G$. For more details, we refer the reader to [2].

## References

1   L. Alvarez-Gaumé, *Supersymmetry and the Atiyah-Singer index theorem*, Commun. Math. Phys. **80** (1983), 161–173.

2   N. Berline, E. Getzler and M. Vergne, "Heat kernels and Dirac operators."

3   E. Getzler, *The degree of the Nicolai map*, J. Func. Anal. **74** (1987), 121–138.

4   S. Kusuoka, *Some remarks on Getzler's degree theorem*, Lecture Notes in Math. **1299** (1988), 239–249.

5   V. Mathai and D. Quillen, *Superconnections, Thom classes and equivariant differential forms*, Topology **25** (1986), 85–110.

6   E. Witten Supersymmetry and Morse theory, J. Differential Geom. **17** (1983), 661–692.

7   E. Witten, *Topological quantum field theory*, Commun. Math. Phys. **117** (1988), 353–386.

# THE GENERAL MATHEMATICAL FRAMEWORK OF EUCLIDEAN QUANTUM MECHANICS, AN OUTLINE

TORBJÖRN KOLSRUD* AND JEAN-CLAUDE ZAMBRINI*,**

* KTH, Stockholm, Sweden
** CFMC, Lisbon, Portugal

Keywords: Quantum Mechanics, Feynman path integrals, Markov processes, Bernstein processes, stochastic calculus of variations, infinite dimensional Ornstein-Uhlenbeck processes.

## Introduction.

In Quantum Physics, the term "Euclidean" means that one, compared to the conventional formulation, works in "imaginary time." It is a general fact that objects associated with Euclidean as opposed to Minkowski space are easier to deal with; this is seen for instance when comparing the usual Laplacian, an elliptic operator, with its Minkowski space counterpart, the hyperbolic d'Alembertian. In Quantum Mechanics, the term "Euclidean" is usually associated with approaches based on the so-called Feynman-Kac formula. The latter originates in attempts to give Feynman's ideas [10, 11] on path integrals a solid mathematical founding.

In this article, Euclidean Quantum Mechanics, EQM henceforth, refers to a distinct and novel Euclidean version of Quantum Mechanics, initiated in 1985-86 [22]. Its principles have been described in a number of articles; see [2, 7, 21] and further references therein.

Roughly speaking, EQM is based on two ideas. The first one is to preserve the Born interpretation of the wave function $\psi$ according to which $|\psi(t, \cdot)|^2$ is a probability density. (Here $\psi(t, \cdot)$ solves the Schrödinger equation with normalised initial data $\phi = \psi(0, \cdot)$.) We can write the probability density as

$$|\psi(t, \cdot)|^2 = \overline{\psi}(t, \cdot) \cdot \psi(t, \cdot) = e^{-itH}\overline{\phi} \cdot e^{itH}\phi \quad (\hbar = 1)$$

where $H$ is the (self-adjoint) Hamiltonian of the system. In EQM the Schrödinger equation and its conjugate are replaced by the corresponding

*Supported by grants from the Swedish Natural Science Research Council, NFR, the Royal Swedish Academy of Sciences, and the Göran Gustafsson Foundation.

diffusion equation and its dual equation. It is fundamental that we have a forward and a backward equation. One is therefore led to investigate a probability density on the form $e^{-tH}\chi^* \cdot e^{tH}\chi$, with $\chi$ and $\chi^*$ carefully chosen. This way we obtain a dynamical theory with the Born interpretation built-in.

The second idea is the basic one in the work of R. Feynman: Quantum Mechanics can be seen as a kind of randomized version of classical mechanics, and the randomness is scaled in units of Planck's constant $\hbar$. The random paths considered are not differentiable. Accordingly, a non-classical calculus has to be used. Such a calculus, directly fitted to the Euclidean approach due to its relation to diffusion processes, was invented by K. Itô almost fifty years ago (see e.g. [13]).

The aim of EQM is first, to carry out a mathematically consistent Euclidean version of Feynman's programme in such a way that–at least at a formal level–formulae derived by him result by analytic continuation back to "real time." That this is possible will be clear from the contents of this and preceeding articles. That it is necessary may be questioned, but the present mathematical status of Quantum Field Theory should not encourage complacency about the probabilistic foundations of the theory. It should be said here, that EQM is fundamentally different from Nelson's "Stochastic Mechanics" (see e.g. [17], and [2, 20] for further comments) in this respect, although some aspects of Stochastic Mechanics have been seminal for the creation of EQM.

The theory of Markov processes constitutes today a highly sophisticated mathematical discipline. It is therefore not entirely surprising that EQM can be extended beyond Feynman's qualitative results. For this very reason, the general mathematical framework of EQM is emphasised in the present article. An effort has been made to sort out the necessary mathematical assumptions in various stages of the construction.

One situation where a broader class of processes–compared to Feynman's formal ones–enters, is in models involving spin degrees of freedom. Let us just mention that it is possible to construct, with the methods described below, Bernstein processes associated with the Pauli equation.

Generally speaking, the presentation is sketchy in that most results and formulae are given without specific assumptions on smoothness, integrability etc. Our aim has been to display the structure of EQM, rather than formulating sharp results. The reader familiar with the relevant parts of probability theory, operator theory and PDE, may consult the standard literature to obtain sufficient conditions. As a consequence there are many interesting mathematical problems left. One is the formulation of EQM in terms of Dirichlet forms; cf. §5.

Rather than sketching the contents of the paper, let us mention two results. In Section 9 we continue the Heisenberg type development of operators from [2]. One can associate operators (observables) on Hilbert space with functionals (processes) on path space so that the corresponding expectations coincide. Defining, for a given Hamiltonian, constants of motion in the usual manner, one proves (Th. 9.7) that these correspond to martingales–the natural path space versions of constants of motion. This dynamical feature is specific to EQM.

Section 10 is devoted to a particular example in infinite dimension, based on the so-called free quantum field. It is indeed possible to create lots of interesting, dynamical, free processes also in infinite dimension. They are free in the sense that a regularised version of the classical (Euclidean) free field equation is satisfied. The reason behind this is the main particularity of our framework, viz. the systematic use of boundary conditions.

It seems rather clear that this example could be extended so as to include interaction terms. A promising candidate for replacing the Gaussian measure considered here, would be the measure appearing in the (two-dimensional) Høegh-Krohn model (see e.g. [19]), corresponding to an exponential interaction of Liouville type.

The main omission in this paper is the stochastic calculus of variations relevant to EQM. This aspect is treated in [7] and a contribution to this volume, [8].

## 1. Basic Assumptions and Notations.

On a nice topological space[1] $M$, for instance $\mathbf{R}^d$, with Borel $\sigma$-algebra $\mathcal{M}$, we consider a sub-Markov transition kernel $p(t, x, dy)$. We write

$$p_t f(x) \equiv \int_M p(t, x, dy) f(y),$$

for the corresponding operator, and assume that $(p_t)_{t>0}$ is a strongly continuous semigroup on $L^2(\mu) \equiv L^2(M, \mathcal{M}, \mu)$ for some Radon measure $\mu \geq 0$ on $(M, \mathcal{M})$, with $\operatorname{supp} \mu = M$. The inner product in $L^2(\mu)$ will be denoted by $(\cdot, \cdot)$.

Mostly we shall consider the case when $(p_t)_{t>0}$ is symmetric on $L^2(\mu)$, and then $p_t = e^{-tH}$ for some self-adjoint operator $H$ on $L^2(\mu)$. There are some applications where non-symmetric cases turn up, for instance in connection with electro-magnetic forces (see [7]). We shall therefore also include the case when $H$ is not self-adjoint but normal on $L^2(\mu)$. The dual semigroup[2] $(p_t^\dagger)_{t>0}$, given by $(p_t f, g) = (f, p_t^\dagger g)$, is then $e^{-tH^\dagger}$. It is not a strongly continuous contraction semigroup.

We shall assume that $p_t$ and $p_t^\dagger$ are associated with Markov processes in the classical manner. Sufficient conditions for this can be found in the standard literature. See e.g. Blumenthal-Getoor [4], Ch. I.9, and Ch. V for processes in duality.

---

[1] To be specific $M$ is locally compact, Hausdorff, and second countable.

[2] Throughout the article we shall see the † to denote adjoint.

## 2. Existence of Markovian Bernstein[3] Processes.

The processes we are about to construct have as parameter space a in general finite interval[4]. For simplicity we shall let this interval be $I \equiv [-1, 1]$. The main difference as compared to the usual construction of a Markov process from a transition function, is that we start from a carefully chosen measure on both endpoints $t = \pm 1$. To accomplish this, let $\theta$, $\theta^* \geq 0$ be measurable functions on $M$ such that

$$(2.1) \qquad \Lambda \rightarrow \int_\Lambda \mu(dx)\theta^*(x)p(2, x, dy)\theta(y)$$

defines a probability measure on $(M \times M, \mathcal{M} \times \mathcal{M})$. (This integral is well defined if $\theta$ and $\theta^*$ are in $L^2$.) Using the semigroup property the right-hand side can be written

$$\int_A \mu(dx)\theta^*(x) \int_M p(1 + t, x, d\xi) \int_C p(1 - t, \xi, dy)\theta(y), \quad t \in I,$$

if $\Lambda = A \times C$. This elementary observation can be used to show that the family of measures $\{\Pi_\Delta\}$, defined for partitions

$$(2.2) \qquad \Delta: -1 = t_0 \leq t_1 \leq t_2 \leq \cdots \leq t_n \leq t_{n+1} = 1,$$

of $I$ by

$$\Pi_\Delta(B_0 \times B_1 \times \cdots \times B_n \times B_{n+1})$$

$$(2.3) \quad = \int \cdots \int_{B_0 \times B_1 \times \cdots \times B_n \times B_{n+1}} \mu(dx_0)\theta^*(x_0) \prod_{i=1}^{n+1} p(t_i - t_{i-1}, x_{i-1}, dx_i)\, \theta(x_{n+1}),$$

where $B_i \in \mathcal{M}$, $0 \leq i \leq n + 1$, form a projective system of probability measures on $(M^I, \mathcal{M}^I)$. We can therefore apply the Daniell-Kolmogorov existence theorem to obtain a process $(Z_t, t \in I)$ with $M$ as state space and with $\{\Pi_\Delta\}$ as finite dimensional distributions.

**(2.4) Theorem.** *There is a process* $(Z_t, t \in I)$, *such that for any partition* $\Delta: -1 \leq t_1 \leq t_2 \leq \cdots \leq t_n \leq 1$ *of* $I$

$$P[Z_{-1} \in B_0, Z_{t_1} \in B_1, \ldots, Z_{t_n} \in B_n, Z_1 \in B_{n+1}]$$

$$= \Pi_\Delta(B_0 \times B_1 \cdots \times B_n \times B_{n+1}),$$

---

[3]Schrödinger was in principle, cf. [18], the initiator of such processes, so the term "Schrödinger processes" could be more appropriate. Bernstein processes are treated in their own right, without reference to quantum dynamics, in Jamison [14]. Beurling's article [3] provides conditions allowing to prescribe the probability densities at both endpoints. For a review of the relations with quantum dynamics, see [2], Ch. 8.

[4]The problem, see Section 3 below, is that in general $e^{tH}$ is not pointwise positive for $t \geq 0$. There are non-trivial situations where one can choose functions $f \geq 0$ such that for all $t$, $e^{tH} f \geq 0$, though, for instance when $H$ is the generator of an Ornstein-Uhlenbeck process in $\mathbf{R}^d$.

*where $\Pi_\Delta$ is defined in (2.3).*

*(2.5) Remark.* Mostly we shall be interested in processes with a.s. continuous paths. For the general case we assume, however, that $t \to Z_t$ is right continuous with finite left-hand limits a.s. Conditions for this (not very restrictive) assumption can be found e.g. in [4].

## 3. The Probability Density.

For a bounded and measurable function $f$ we can write the expectation of $f(Z_t)$ as

$$(3.1) \qquad \mathbf{E}[f(Z_t)] = \int f \, \rho(t) d\mu$$

where $\rho(t) = \rho(t, \cdot)$ is the probability density of $Z_t$. To see this, and to get an expression for $\rho$, note that the left-hand side equals

$$(\theta^*, p_{1+t}(f p_{1-t}\theta)) = (p_{1+t}^\dagger \theta^*, f p_{1-t}\theta) = (f, p_{1+t}^\dagger \theta^* \cdot p_{1-t}\theta) \equiv (f, \rho(t)).$$

We shall use the notation

$$\theta^*(t) = \theta^*(t, \cdot) \equiv p_{1+t}^\dagger \theta^*, \quad t \geq -1,$$

and

$$\theta(t) = \theta(t, \cdot) \equiv p_{1-t}\theta, \quad t \leq 1.$$

The definitions are made so that $\theta^* = \theta^*(-1)$ and $\theta = \theta(1)$. Defining $\chi \equiv \theta(0)$ and $\chi^* \equiv \theta^*(0)$, we could also write

$$\theta^*(t) = e^{-(1+t)H^\dagger}\theta^* = e^{-tH^\dagger}\chi^*, \quad \theta(t) = e^{-(1-t)H}\theta = e^{tH}\chi,$$

and then

$$(3.2) \qquad \rho(t) = \theta^*(t)\theta(t) = e^{-tH^\dagger}\chi^* e^{tH}\chi, \quad |t| \leq 1.$$

This expression displays the dynamical content of the construction. If $\theta \in \mathrm{Dom}(H)$ and $\theta^* \in \mathrm{Dom}(H^\dagger)$, we get

$$\frac{\partial}{\partial t}\rho(t) = \left(-H^\dagger e^{-tH^\dagger}\chi^*\right) \cdot e^{tH}\eta + e^{-tH^\dagger}\chi^*\left(He^{tH}\chi\right),$$

so

$$\frac{d}{dt}\int \rho(t)\, d\mu = (-H^\dagger e^{-tH^\dagger}\chi^*, e^{tH}\chi) + (e^{-tH^\dagger}\chi^*, He^{tH}\chi) = 0.$$

This is the Euclidean version of the famous probabilistic interpretation, due to Born, of quantum mechanics. The quantum mechanical law of conservation of probability is (cf. the introduction)

$$\frac{d}{dt}\int \overline{\psi}(t)\psi(t)\,dx = 0,$$

where $\psi(t)$ solves the Schrödinger equation. The Euclidean version of the Born interpretation appeared (in a special case) in Schrödinger's 1932 paper [18].

*(3.3) Remark.* We note that in the self-adjoint case $(H = H^\dagger)$ the time reflection $t \to -t$ corresponds to letting $(\theta^*,\theta) \to (\theta,\theta^*)$. In particular $Z_t$ and $Z_{-t}$ are equally distributed if $\theta^* = \theta$, since then $\rho(-t) = e^{(1-t)H}\theta e^{(1+t)H}\theta = \rho(t)$. Together with the Markov property of $Z$, shown in the next section, reflection positivity ([19]) follows in this case.

**4. Markov Property. Forward and Backward Propagators.**
We first note the following simple consequence of the definition of $\theta(t)$:

(4.1)             $\theta(s) = e^{-(t-s)H}\theta(t), \quad -1 \le s \le t \le 1.$

Consequently

(4.2)     $q(s,x;t,dy) \equiv \frac{1}{\theta(s,x)}p(t-s,x,dy)\theta(t,y), \quad -1 \le s \le t \le 1,$

is a probability measure on $M$ for all $x$ such that $\theta(s,x) \neq 0$. This holds for $\mu$-a.e. $x$ if $p_t$ is *positivity improving*, i.e.

(4.3)             $0 \le f,g \in L^2, \quad f,g \not\equiv 0 \implies (p_t f,g) > 0.$

We assume (4.3) from now on, and we write $q_{s,t}$ for the operator corresponding to (4.2), i.e.

(4.4)         $q_{s,t}f = \int q(s,\cdot;t,dy)f(y) = \frac{1}{\theta(s)}p_{t-s}(f\theta(t)).$

It is easily seen that the $q_{s,t}$ are Markovian (and non-stationary in general) in that

(4.5)             $q_{s,t} \circ q_{t,u} = q_{s,u}, \quad -1 \le s \le t \le u \le 1.$

Furthermore, the positivity improving property (4.3) implies that for a partition of $I$ as in (2.2) we can write
(4.6)
$$\mathbf{E}[f_1(Z_{t_1})\cdots f_n(Z_{t_n})] = (\rho(-1), q_{t_0,t_1}(f_1 q_{t_1,t_2}(f_2\cdots q_{t_{n-1},t_n}f_n))\cdots))),$$

which means that $Z$ is a non-stationary Markov process indexed by $I$ and with initial distribution $\rho(-1)$. Equivalently, we have obtained a *forward* Markovian description of $Z$ by conjugating with a solution $\theta(\cdot)$ of the *backward* heat equation obtained from $H$.

There is another description of $Z$ as a *backward* Markov process, and this is obtained by conjugating with a solution $\theta^*(\cdot)$ of the *forward* dual heat equation. Instead of (4.1) we use the relation

$$(4.1^*) \qquad \theta^*(t) = e^{-(t-s)H^\dagger}\theta^*(s), \quad -1 \le s \le t \le 1,$$

to get the kernel

$$(4.2^*) \quad q^*(t, dx; s, y) \equiv \frac{1}{\theta^*(t, y)} p^\dagger(t - s, y, dx)\theta^*(s, x), \quad -1 \le s \le t \le 1,$$

and the corresponding operator

$$(4.4^*) \qquad q^*_{t,s} f = \int f(x) q^*(t, dx; s, \cdot) = \frac{1}{\theta^*(t)} p^\dagger_{t-s}(f\theta^*(s)).$$

It is clear from (4.3) that $p_t$ and $p_t^\dagger$ are positivity improving simultaneously.

The formulae corresponding to (4.5) and (4.6) are

$$(4.5^*) \qquad q^*_{u,t} \circ q^*_{t,s} = q^*_{u,s}, \quad -1 \le s \le t \le u \le 1,$$

and
$(4.6^*)$
$$\mathbf{E}\big[f_n(Z_{t_n}) \cdots f_1(Z_{t_1})\big] = \big(q^*_{t_n,t_{n-1}}(f_n \cdots q^*_{t_2,t_1} f_2(q^*_{t_1,t_0} f_1)) \cdots \big), \rho(1)\big).$$

Here we look upon $Z$ as starting at $t = 1$ according to the distribution $\rho(1)$ and developing backwards in time according to $q^*_{t,s}$. The first picture corresponds to the increasing family of $\sigma$-algebras

$$\mathcal{F}_t = \sigma\{Z_s, -1 \le s \le t\}, \quad t \in I,$$

and the second to the decreasing family

$$\mathcal{F}_t^* = \sigma\{Z_s, t \le s \le 1\}, \quad t \in I.$$

*(4.7) Remarks.* 1. The assumption (4.3) is really an irreducibility condition related to ergodicity. See [1], pp. 22-24, for a discussion of related matters. 2. Independently of Schrödinger's original motivation, i.e. the probabilistic foundations of quantum mechanics mentioned in Section 2, the present class of processes is an interesting subject for purely probabilistic investigations. See e.g. Wakolbinger [20], and further references therein.

## 5. Generators. Forward and Backward Derivatives.

We shall now compute the generators of $Z$ w.r.t. the forward and backward descriptions. By the Markov property it suffices to condition w.r.t. the present. With $\mathbf{E}_t \equiv \mathbf{E}[\cdot | Z_t]$ we therefore define, for sufficiently regular functions $f = f(t, x)$,

$$(5.1) \qquad Df(t, Z_t) = \lim_{h \downarrow 0} \frac{1}{h} \mathbf{E}_t[f(t+h, Z_{t+h}) - f(t, Z_t)].$$

We write $Df(t, x)$ for $Df(t, Z_t)$ conditioned on $Z_t = x$. Then

$$(5.2) \qquad Df(t, \cdot) = \lim_{h \downarrow 0} \frac{1}{h} \left( q_{t, t+h} f(t+h, \cdot) - q_{t,t} f(t, \cdot) \right),$$

and a simple calculation yields

$$(5.3) \qquad Df(t, \cdot) = \frac{1}{\theta(t)} \left( \frac{\partial}{\partial t} - H \right)(f\theta(t)).$$

It is convenient to single out the 0-order term of $H$ and write $H = H_0 + V$, where $H_0 1 = 0$. Defining

$$(5.4) \qquad \Gamma(f, g) \equiv H_0 f \cdot g + f H_0 g - H_0(fg)$$

we obtain a pointwise bilinear and symmetric form such that

$$(5.5) \qquad Df(t, \cdot) = \frac{\partial f}{\partial t} - H_0 f + \frac{1}{\theta(t)} \Gamma(f, \theta(t)).$$

The derivative $D_*$ obtained by conditioning in the future is defined by

$$(5.6) \qquad \begin{aligned} D_* f(t, \cdot) &= \lim_{h \downarrow 0} \frac{1}{h} \mathbf{E}_t[f(t, Z_t) - f(t-h, Z_{t-h})] \\ &= \lim_{h \downarrow 0} \frac{1}{h} \left( q^*_{t,t} f(t, \cdot) - q^*_{t,t-h} f(t-h, \cdot) \right), \end{aligned}$$

and this time one finds

$$(5.7) \qquad \begin{aligned} D_* f(t, \cdot) &= \frac{1}{\theta^*(t)} \left( \frac{\partial}{\partial t} + H^\dagger \right)(f\theta^*(t)) \\ &= \frac{\partial f}{\partial t} + H_0^\dagger f - \frac{1}{\theta^*(t)} \Gamma^*(f, \theta^*(t)), \end{aligned}$$

where $\Gamma^*$ is defined in analogy with $\Gamma$ in (5.4), but $H$ is replaced by $H^\dagger$.

*(5.8) Examples.* (a) When $M = \mathbf{R}^d$ and $H_0 = -\frac{1}{2}\Delta + \phi \cdot \nabla$, $\Gamma(f_1, f_2) = \nabla f_1 \cdot \nabla f_2$. More generally we get $\Gamma(f_1, f_2) = \sum g^{ij} \partial_i f_1 \partial_j f_2$ when $\Delta$ is the Laplace-Beltrami operator in a Riemannian manifold $(M, g)$ and then

$\phi \cdot \nabla$ is interpreted as a vector field. Note that these expressions hold independently of the drift $\phi$ and the potential $V$.

(b) Consider now the case where $M = \mathbf{R}^d$ and $H$ the self-adjoint differential operator

$$(5.9) \qquad H = H^\dagger = -\frac{1}{2}\Delta + \phi \cdot \nabla + V \equiv H_0 + V.$$

For self-adjointness we need $\phi = \nabla\Phi$ and then $d\mu/dx = e^{-2\Phi}$. We assume that $\phi$ and $V$ are sufficiently smooth to allow for the calculations to follow. $\Gamma$ was calculated in (a), and one finds

$$(5.10) \qquad Df(t, \cdot) = \frac{\partial f}{\partial t} - H_0 f + \beta(t, \cdot) \cdot \nabla f,$$

where

$$(5.11) \qquad \beta(t) = \beta(t, \cdot) \equiv \frac{\nabla\theta}{\theta}(t).$$

Similarly

$$(5.10^*) \qquad D_* f(t, \cdot) = \frac{\partial f}{\partial t} + H_0 f - \beta^*(t, \cdot) \cdot \nabla f,$$

where

$$(5.11^*) \qquad \beta^*(t) \equiv -\frac{\nabla\theta^*}{\theta^*}(t).$$

(c) Under suitable conditions on closability, see [12], we can express $Df$ using a family of Dirichlet energy forms indexed by $I^0$, the interior of $I$. For $f$ independent of time, we have

$$(5.12) \qquad \frac{1}{2}\int_M \nabla f \cdot \nabla g \, \theta(t)^2 \, d\mu = -\int_M Df \cdot g \, \theta(t)^2 \, d\mu,$$

with a similar expression for $D_*$.

We see that in this case the theory could be formulated via two families of symmetric diffusions[5], each indexed by $t \in I^0$.

(d) In general $\Gamma$ is non-local. The following generic discrete example shows that it is a kind of inner product between generalised gradients. We have $M = \mathbf{Z}^d$ with $\mu$ the counting measure and

$$Hf(x) = H_0 f(x) = f(x) - \sum_{|x-y|\geq 1} a(x,y)f(y), \qquad x \in \mathbf{Z}^d,$$

where $a(x,y) \geq 0$ and $\sum_{y\neq x} a(x,y) = 1$ for all $x$. A simple calculation shows that

$$\Gamma(f,g) = \sum_y a(x,y)(f - f(y))(g - g(y)).$$

---

[5]For such processes and their relation to Dirichlet spaces, see [12].

## 6. Duality. Generalised h-transform.

Our starting point is the easily proved formula

$$(6.1) \qquad \int f q_{s,t} \cdot g \, \rho(s) d\mu = \int q_{t,s}^* f \cdot g \, \rho(t) d\mu,$$

indicating that an infinitesimal version should lead to some kind of duality between $D$ and $D_*$. Let us introduce the (space-time) Hilbert space $\mathcal{H} = L^2(dt \otimes d\mu, I \times M) = L^2(\nu)$. With[6] $\hat{\mathcal{H}} = L^2(\theta^2 d\nu)$, the map

$$\mathcal{H} \ni f \to \hat{f} \equiv \frac{f}{\theta} \in \hat{\mathcal{H}},$$

is unitary, and similarly for

$$\mathcal{H} \ni g \to \hat{g}^* \equiv \frac{g}{\theta^*} \in \hat{\mathcal{H}}^*,$$

where $\hat{\mathcal{H}}^* = L^2(\theta^{*2} d\nu)$. Any operator $A$ on $\mathcal{H}$ carries over to operators on $\hat{\mathcal{H}}$ and $\hat{\mathcal{H}}^*$ by conjugation. That is

$$\hat{A}\phi = \theta^{-1} A(\phi\theta), \quad \phi \in \hat{\mathcal{H}}, \qquad \hat{A}^*\psi = \theta^{*-1} A(\psi\theta^*), \quad \psi \in \hat{\mathcal{H}}^*.$$

The measure $\rho d\nu$ gives rise to a pairing $_{\hat{\mathcal{H}}}\langle \cdot, \cdot \rangle_{\hat{\mathcal{H}}^*}$ between $\hat{\mathcal{H}}$ and $\hat{\mathcal{H}}^*$, and this leads to a kind of unitary equivalence with $\mathcal{H}$, carried over also to operators. We have

$$_{\hat{\mathcal{H}}}\langle \hat{A}\hat{f}, \hat{g}^* \rangle_{\hat{\mathcal{H}}^*} \equiv \langle \hat{A}\hat{f}, \hat{g}^* \rangle \equiv \int \hat{A}\hat{f} \cdot \hat{g}^* \, \rho d\nu = \int A\left(\theta \cdot \frac{f}{\theta}\right) \cdot g \, d\nu = (Af, g)_{\mathcal{H}}.$$

We see that with respect to this pairing

$$(6.2) \qquad (A^\dagger)\hat{} = (\hat{A})^*.$$

Consider now the operators $D = (\partial/\partial t - H)\hat{}$ and $D_* = (\partial/\partial t + H^\dagger)\hat{}^*$ with Dirichlet boundary conditions at $t = \pm 1$. This means that they are densely defined on the images of $C_0^\infty(I) \otimes \text{Dom}(H)$ and $C_0^\infty(I) \otimes \text{Dom}(H^\dagger)$ respectively. More precisely

$$\text{Dom}(D) = \left\{ \frac{f}{\theta} : f \in C_0^\infty(I) \otimes \text{Dom}(H) \right\},$$

with a similar expression for $D_*$.

---

[6] In the rest of this section we only consider functions of $t$ and $x$, and to simplify the notation we shall write $\theta$ for $\theta(t)$ and $\rho$ for $\rho(t)$ etc.

Since $\partial/\partial t$ and $-\partial/\partial t$ are adjoint under Dirichlet boundary conditions, we deduce from (6.2) that with the mentioned domains of $D$ and $D_*$ we have $D_* = -D^\dagger$:

$$\langle D\phi, \psi \rangle = -\langle \phi, D_*\psi \rangle, \quad \phi \in \hat{\mathcal{H}}, \ \psi \in \hat{\mathcal{H}}^*.$$

This is the infinitesimal version of (6.1) with Dirichlet boundary conditions in the time direction. In general this holds modulo boundary terms.

*(6.3) Remark.* The material developed in this section is a space-time extension of a well-known transformation. In its probabilistic manifestation it is known as Doob's $h$-transform, whereas mathematical physicists call it the ground state transformation.

## 7. The Stochastic Equations of Motion.

In this section we shall exclusively consider the case where $M = \mathbf{R}^d$ and $H$ is a self-adjoint differential operator, as in (5.9) ff. With $f(t, x) = x$ we obtain from Eq. (5.10) the *regularised forward velocity*
(7.1)

$$DZ_t = \lim_{h\downarrow 0} \mathbf{E}\Big[\frac{1}{h}(Z_{t+h} - Z_t)\big|\mathcal{F}_t\Big] = \beta(t, Z_t) - \phi(Z_t) = \frac{\nabla\theta}{\theta}(t, Z_t) - \phi(Z_t),$$

and from Eq. (5.10*) the *regularised backward velocity*
(7.2)

$$D_*Z_t = \lim_{h\downarrow 0} \mathbf{E}\Big[\frac{1}{h}(Z_t-Z_{t-h})\big|\mathcal{F}_t^*\Big] = \beta^*(t, Z_t)+\phi(Z_t) = -\frac{\nabla\theta^*}{\theta^*}(t, Z_t)+\phi(Z_t),$$

cf. (5.11) and (5.11*).

Writing momentarily $\theta$ for $\theta(t)$ etc., we have

$$\Big(\frac{\partial}{\partial t} - H_0\Big)\beta = \Big(\frac{\partial}{\partial t} + \tfrac{1}{2}\Delta - \phi\cdot\nabla\Big)\frac{\nabla\theta}{\theta}$$

$$=\frac{\dot{\nabla}\theta}{\theta} - \frac{\dot{\theta}\nabla\theta}{\theta^2} + \tfrac{1}{2}\frac{\nabla\Delta\theta}{\theta} - \tfrac{1}{2}\frac{\nabla\theta}{\theta}\cdot\frac{\Delta\theta}{\theta} - \tfrac{1}{2}\nabla\Big(\frac{\nabla\theta}{\theta}\Big)^2 - \phi\frac{\Delta\theta}{\theta} + \phi\Big(\frac{\nabla\theta}{\theta}\Big)^2$$

$$=\frac{1}{\theta}(\nabla - \beta)(\dot{\theta} + \tfrac{1}{2}\Delta\theta) - \tfrac{1}{2}\nabla\beta^2 - \phi\frac{\nabla\theta}{\theta} + \phi\beta^2$$

$$=\frac{1}{\theta}(\nabla - \beta)(\phi\nabla\theta + V\theta) - \beta\nabla\beta - \phi\frac{\nabla\theta}{\theta} + \phi\beta^2 = \nabla V - \beta\nabla(\phi - \beta).$$

Now

$$DDZ = D(\beta - \phi) = \Big(\frac{\partial}{\partial t} - H_0\Big)(\beta - \phi) + \beta\nabla(\beta - \phi)$$

$$= \Big(\frac{\partial}{\partial t} - H_0\Big)\beta + H_0\phi + \beta\nabla(\beta - \phi),$$

so we obtain the *stochastic forward equation of motion*

(7.3) $$DDZ = \nabla V(Z) + H_0\phi(Z).$$

One proves in very much the same manner that this result also holds for the decreasing filtration $(\mathcal{F}_t^*)$. The *stochastic backward equation of motion* is therefore

$$(7.4) \qquad D_* D_* Z = \nabla V(Z) + H_0 \phi(Z).$$

Two extreme cases turn up: no drift, and no potential. In the first case we get the classically looking Euclidean, regularised, Newton equations

$$(7.5) \qquad DDZ = D_* D_* Z = \nabla V(Z),$$

cf. [2, 7, 21]. In the second case,

$$(7.6) \qquad DDZ = D_* D_* Z = H_0 \phi(Z).$$

Now

$$H_0 \phi = \tfrac{1}{2} \Delta \phi - \phi \nabla \phi = -\tfrac{1}{2} \Delta \nabla \Phi - \nabla \Phi \nabla \otimes \nabla \Phi = -\tfrac{1}{2} \nabla \big( \Delta \Phi - |\nabla \Phi|^2 \big),$$

and

$$\big( -\tfrac{1}{2} \Delta + \tilde{V} \big) e^{-\Phi} = 0,$$

if $\tilde{V}$ is defined as

$$\tilde{V} \equiv -\tfrac{1}{2} \big( \Delta \Phi - |\nabla \Phi|^2 \big).$$

Similarly to the reasoning in §6, one sees that the operators $-\tfrac{1}{2}\Delta + \tilde{V}$ and $-\tfrac{1}{2}\Delta + \phi \cdot \nabla$ are unitarily equivalent under conjugation with $e^{\pm \Phi}$. Thus the stochastic equations of motion are in the natural sense invariant under the unitary maps $f \to f e^{\pm \Phi}$. The dynamics for $Z$ remains the same and the interaction is always represented on potential form. The total potential is

$$(7.7) \qquad W \equiv V + \tilde{V} = V - \tfrac{1}{2} \big( \Delta \Phi - |\nabla \Phi|^2 \big).$$

*(7.8) Remarks.* The mentioned unitary equivalence is easily seen from another point of view in Section 9 below. Furthermore, we mention the well-known fact that the passage between a drift term (in our case $\phi \cdot \nabla$) and a potential (in our case $\tilde{V}$) can also be accomplished by use of Girsanov's formula.

## 8. Path Integral Representations.

We saw in the preceeding section that $DDZ = \nabla W(Z)$ where $W = V + \tfrac{1}{2}(|\nabla \Phi|^2 - \Delta \Phi)$ is the (total) potential. One might ask whether this is actually an Euler-Lagrange equation. At the classical limit, where $D$ and $D_*$ coincide with the usual time-derivative, the associated Euclidean Lagrangian would be $L = \tfrac{1}{2}|\dot{Z}|^2 + W(Z)$. Let us show that this remains true here if $\dot{Z}$ is replaced by the regularised derivative $DZ$, and then that the dynamical structure is preserved. This holds for the backward description too.

To express the "Euclidean backward wave function" $\theta(t)$ as a path integral we let the forward (!) action density be

$$(8.1) \qquad \alpha(t) \equiv -\log\theta(t) + \Phi,$$

so that $\alpha$ is the integral of the forward velocity $DZ$. One finds

$$-D\alpha(t) = V + \tfrac{1}{2}(\beta(t) - \phi)^2 + \frac{\phi^2}{2} - \tfrac{1}{2}\nabla\cdot\phi = \tfrac{1}{2}(\beta(t) - \phi)^2 + W,$$

so

$$(8.2) \qquad D\alpha(t, Z_t) = -\tfrac{1}{2}(DZ_t)^2 - W(Z_t) \equiv -L(DZ_t, Z_t),$$

with $L$ denoting the *classical* Euclidean Lagrangian. By Dynkin's formula

$$\mathbf{E}_t[\alpha(1, Z_1)] - \alpha(t, Z_t) = \mathbf{E}_t \int_t^1 D\alpha(s, Z_s)\,ds = -\mathbf{E}_t \int_t^1 L(DZ_s, Z_s)\,ds,$$

which upon exponentiating leads to the path integral formula

$$(8.3) \qquad \theta(t, x)e^{-\Phi(x)} = \exp-\left\{ \mathbf{E}_t^x \int_t^1 L(DZ_s, Z_s)\,ds + \mathbf{E}_t^x[\alpha(1, Z_1)] \right\},$$

where $\mathbf{E}_t^x$ is the expectiation conditioned on $Z_t = x$.

Using the action density

$$(8.1^*) \qquad \alpha^*(t) \equiv -\log\theta^*(t) + \Phi$$

for the backward description, we similarly get
$$(8.3^*)$$

$$\theta^*(t, x)e^{-\Phi(x)} = \exp-\left\{ \mathbf{E}_t^x \int_{-1}^t L(D_*Z_s, Z_s)\,ds + \mathbf{E}_t^x[\alpha^*(-1, Z_{-1})] \right\}.$$

This suggests also that the forward and backward equations of motion in Section 7 are Euler-Lagrange equations for the pair of regularised action functionals involved in the path integral representations. The calculus of variations necessary to prove this is presented in [8], and its qualitative conclusions coincide with those of Feynman's.

## 9. Operator (Observable) and Path Space Calculus.

We start by deriving a Euclidean version of the Heisenberg picture of Quantum Mechanics. If $A$ is an operator (observable) on our basic Hilbert space $L^2(\mu)$, we obtain two families $A(t)$ and $A^*(t)$ of operators, indexed by $t \in I$, corresponding to the forward and backward developments. We shall only consider the case where the given Hamiltonian $H$ is self-adjoint, but the extension to more general cases is straight forward.

To avoid problems with the domains of definition, we shall assume that the operators in question can act invariantly on a set of vectors that are analytic for $H$. We refer to [2] for details.

As in Section 3 we let $\chi = \theta(0)$ and $\chi^* = \theta^*(0)$, so that $\theta(t) = e^{tH}\chi$ and $\theta^*(t) = e^{-tH}\chi^*$. With

$$(9.1) \qquad A(t) = A_H(t) = e^{-tH} A e^{tH}, \quad t \in I,$$

one easily finds
$$(9.2)$$
$$(\theta^*(t), A\theta(t)) = (\theta^*(t), A(0)\theta(t)) = (\theta^*(0), A(t)\theta(0)) = (\chi^*, A(t)\chi), \quad t \in I,$$

and, similarly,

$$(9.2^*) \qquad (B\theta^*(t), \theta(t)) = (B^*(t)\chi^*, \chi), \quad t \in I,$$

where we have defined

$$(9.1^*) \qquad B^*(t) = B^*_H(t) = e^{tH} B e^{-tH} = B(-t), \quad t \in I.$$

We have
$$(\theta^*(t), A\theta(t)) = (A^\dagger \theta^*(t), \theta(t)),$$

so
$$(A^\dagger)^*(t) = A^\dagger(-t) = (A(t))^\dagger.$$

The notation $B^*(t)$ may seem a bit superflous in this set-up, but it becomes more natural when considering a family of forward and backward Hilbert spaces, as in ref. [2].

Let $R_A$ and $R^*_B$ denote the functions

$$(9.3) \qquad R_A(t,x) = \frac{A\theta}{\theta}(t,x), \quad R^*_B(t,x) = \frac{B\theta^*}{\theta^*}(t,x), \quad t \in I, \, x \in M.$$

We consider the ordered pair $(\chi^*, \chi)$ as a state, and define the Euclidean expectation of $A(t)$ in this state by

$$(9.4) \qquad \langle A(t) \rangle \equiv \langle A(t) \rangle_{\chi^*, \chi} \equiv (\chi^*, A(t)\chi).$$

Similarly

$$(9.4^*) \qquad \langle B^*(t) \rangle \equiv \langle B^*(t) \rangle_{\chi^*, \chi} \equiv (B^*(t)\chi^*, \chi).$$

The operator expectations can also be expressed as expectations over path space:
$$\langle A(t) \rangle = \mathbf{E}\{R_A(t, Z_t)\}, \quad \langle B^*(t) \rangle = \mathbf{E}\{R^*_B(t, Z_t)\}.$$

The construction is made so that

(9.5)
$$\frac{d}{dt}A(t) = [A, H](t),$$

and

(9.5*)
$$\frac{d}{dt}B^*(t) = [H, B]^*(t).$$

Hence, under the correspondences $A \to R_A$ and $B \to R_B^*$

$$\frac{d}{dt}A(t) \longrightarrow R_{[A,H]}(t, \cdot), \qquad \frac{d}{dt}B^*(t) \longrightarrow R_{[H,B]}^*(t, \cdot).$$

We now introduce the natural Euclidean analogues of constant quantum mechanical observables.

*(9.6) Definition.* We say that the operator $A$ is a *forward constant of motion* if $dA(t)/dt \equiv 0$. Similarly, $B$ is a *backward constant of motion* if $dB^*(t)/dt \equiv 0$.

The following theorem gives an amplified version of quantum conservation laws[7]:

**(9.7) Theorem.** *For any forward (backward) constant of motion $A$ ($B$), the process $\{R_A(t, Z_t), t \in I\}$ ( $\{R_B^*(t, Z_t), t \in I\}$ ) is an $\{\mathcal{F}_t, t \in I\}$- ( $\{\mathcal{F}_t^*, t \in I\}$-)martingale.*

*Proof.* We first note that $A$ is a constant of motion if and only if $A$ and $H$ commutes. We note next that

(9.8)
$$R_{[A,H]} = DR_A,$$

so $DR_A = 0$, which implies that $R_A(t, Z_t)$ is a martingale with respect to the forward filtration $(\mathcal{F}_t, t \in I)$.

To prove (9.8) we recall from (5.3) that $Df(t, \cdot) = (\partial/\partial t - H)(f\theta(t))/\theta(t)$, whence

$$DR_A(t, \cdot) = \frac{1}{\theta(t)}\left(\frac{\partial}{\partial t} - H\right)(A\theta(t)) = \frac{1}{\theta(t)}(A\dot\theta(t)) - HA\theta(t))$$

$$= \frac{1}{\theta(t)}(AH\theta(t)) - HA\theta(t)) = \frac{1}{\theta(t)}[A, H]\theta(t) = R_{[A,H]}(t, \cdot).$$

*(9.9) Remarks.* (a) The condition that $A$ and $H$ commute is not as strong as it seems. It does *not* mean that $A$ commutes with the spectral family

---

[7]In spite of a remark in [17], the second author did not find, in the context of Nelson's theory, this kind of strong, i.e. pointwise, quantum conservation laws

associated with $H$, rather it is a statement on invariant subspaces. For instance, if $H = \sum \lambda_n P_n$, where the mutually orthogonal projections $P_n$ all have finite dimensional ranges $V_n$, say, then any $A$ on the form $A = \bigotimes A_n$, with $A_n \in gl(V_n)$, commutes with $H$.

(b) It is perhaps a bit surprising that Theorem (9.7) holds without assuming that the Hamiltonian has some particular form. However, when $H$ is as in (5.9), and $M = \mathbf{R}^d$, we can say more. Define the operators $Q^i$, $1 \leq i \leq d$, as multiplication with the co-ordinates $x^i$, and $P_j$ as $\partial_j - \phi_j$, $1 \leq j \leq d$. The calculations of Section 7 really displays the usual, but Euclidean, equations of motion for position and momenta, i.e.

$$(9.10) \qquad \dot{Q}^i = P_i, \quad \dot{P}_i = \nabla_i W(Q),$$

cf. [2, 21].

A more interesting relation is the following. With the action density $\alpha(t) = -\log \theta(t) + \Phi$ as in Eq. (8.1), and with $d$ denoting the total differential w.r.t. $t$ and $x$, we have

$$(9.11) \qquad -d\alpha = R_H dt + \sum R_{P_i} dx^i \equiv E dt + p \cdot dx,$$

where $E$ denotes energy and $p$ momentum as functions of $(t, x)$. It is interesting and satisfactory to note that this one-form, fundamental in symplectic geometry, turns up naturally in EQM.

(c) As a last application we derive the Heisenberg commutation relations for the general self-adjoint Hamiltonian in (5.9). It is more natural to work directly on path space. Clearly $[Q^i, P_j]$ should correspond to an infinitesimal version of

$$(9.12) \qquad \frac{1}{h}\left( Z_t^i(Z_t^j - Z_{t-h}^j) - (Z_{t+h}^j - Z_t^j)Z_t^i \right), \quad h > 0.$$

Letting $h \downarrow 0$ and taking expectations we get

$$\mathbf{E}\{Z_t^i(D_* Z_t^j - D Z_t^j)\} = \int x^i(\beta^*(t)_j - \beta_j(t) + 2\phi_j)\,\rho(t)d\mu$$

$$(9.13) \qquad = \int x^i\left( -\frac{\partial_j \theta^*}{\theta^*}(t) - \frac{\partial_j \theta}{\theta}(t) + 2\phi_j \right)\theta\theta^*(t)e^{-2\Phi}\,dx$$

$$= -\int x^i \partial_j(\theta\theta^*(t)e^{-2\Phi})\,dx$$

$$= \int (\partial_j x^i)\,\theta\theta^*(t)e^{-2\Phi}\,dx = \delta_j^i \int \rho(t)d\mu = \delta_j^i.$$

This is the Euclidean version of a crucial result of Feynman's. Cf. [10, 11] and [7].

**10. An Infinite Dimensional Example.**
In this section we shall study an example based on a certain infinite dimensional Ornstein-Uhlenbeck semigroup related to the so-called free quantum field. (Cf. [15, 17, 19].)

Denote by $\Lambda$ the self-adjoint operator $\Lambda \equiv \sqrt{-\Delta + m^2}$, where $\Delta$ is the Laplacian in $\mathbf{R}^d$, and $m > 0$. Denote by $\mathcal{H}^{1/2}$ the Sobolev space with inner product $\int \xi \Lambda \eta$, and by $\mathcal{H}^{-1/2}$ its dual. We let $\mu$ be the centred Gaussian measure that can be realised on the space of Schwartz distributions $\mathcal{S}'(\mathbf{R}^d)$, with covariance given by the inner product in $\mathcal{H}^{-1/2}$. In terms of its Fourier transform, $\mu$ is defined by

$$(10.1) \qquad \int_{\mathcal{S}'(\mathbf{R}^d)} e^{i\langle\cdot,\xi\rangle}\, d\mu = e^{-\frac{1}{2}\int \xi \Lambda^{-1}\xi}, \quad \xi \in \mathcal{S}(\mathbf{R}^d).$$

We can choose a Hilbert space $M \supset \mathcal{H}^{-1/2}$ such that $\mu(M) = 1$, simply by arranging that the inclusion $\mathcal{H}^{-1/2} \subset M$ is a Hilbert-Schmidt mapping. The relation between the spaces mentioned so far is

$$(10.2) \qquad \mathcal{S}(\mathbf{R}^d) \subset \mathcal{H}^{1/2} \subset L^2(\mathbf{R}^d) \subset \mathcal{H}^{-1/2} \subset M \subset \mathcal{S}'(\mathbf{R}^d).$$

The O-U semigroup $(p_t)$ and its generator $H$ is most simply described in Fock space terms (see [19]) as

$$(10.3) \qquad H = d\mathrm{Exp}\,(\Lambda), \quad p_t = e^{-tH} = \mathrm{Exp}\,(e^{-t\Lambda}), \ t \geq 0,$$

where $\mathrm{Exp}\,(e^{-t\Lambda})$ indicates $e^{-t\Lambda}$ acting component-wise on each factor of elements in the symmetric tensor algebra over $L^2(\mathbf{R}^d)$. $H$ is also implicitly defined through the relation $\int f H g\, d\mu = \frac{1}{2}\int (\nabla f | \nabla g)\, d\mu$, where the nablas and the inner product refers to $\mathcal{H}^{1/2}$. Below, $X = (X_t, t \geq 0)$ will denote the corresponding $M$-valued O-U process, and $\mathbf{E}^X$ its expectation operator.

We can now start to construct infinite dimensional Bernstein processes. First, if $\theta, \theta^* \geq 0$ are non-zero and suitably normalised elements of $L^2(\mu)$, then (2.1) makes sense, and defines a probability measure on $M \times M$. Again, (2.3) defines a projective system of probability measures. The Daniell-Kolmogorov theorem is valid for complete and separable metric spaces, see Dellacherie-Meyer [9], Ch. III.50-53, pp. 109-114. Hence Theorem 2.4 extends to the case under consideration, and we obtain an $M$-valued Bernstein process $Z = (Z_t, t \in I)$ whose finite dimensional distributions are given by Eq. (2.3).

Clearly Section 3 carries over directly. In particular, the Born interpretation holds. It is a general result that O-U semigroups are positivity improving, see e.g. Simon [19], Theorem I.12. Hence the arguments in Section 4 carry over to show that $Z$ is doubly Markovian with forward and backward descriptions provided by Eqs. (4.4-6) and (4.4*-6*), respectively.

Since the process $X$ is continuous (see e.g. [15]), one expects that this holds for $Z$ too. The following result gives an affirmative answer to this for a wide class of functions $\theta, \theta^*$.

**(10.4) Theorem.** *Assume that $\theta$ and $\theta^*$ are in $L^4(\mu)$. Then $I \ni t \to Z_t \in M$ is a.s. Hölder continuous of any order $< 1/2$.*

*Proof.* Let $f \geq 0$ be an $\mathcal{M} \otimes \mathcal{M}$-measurable function, and fix $0 \leq s \leq t$. Denote by $\nu(dxdy)$ the measure $p(t - s, x, dy)\mu(dx)$ on $M \times M$. Hölder's inequality gives

(10.5)
$$\mathbf{E}[f(Z_s, Z_t)] = \int f(x, y) \, \theta^*(s, x)\theta(t, y)\nu(dxdy)$$
$$\leq \left\{ \int f^2 \, d\nu \right\}^{1/2} \left\{ \int \theta^*(s)^2 \otimes \theta(t)^2 \, d\nu \right\}^{1/2}.$$

Now, again by Hölder's inequality, and independently of $s$ and $t$,
(10.6)
$$\left\{ \int \theta^*(s)^2 \otimes \theta(t)^2 \, d\nu \right\}^{1/2} = \|\theta^*(s) \otimes \theta(t)\|_{2,\nu} \leq \|p_{t-s}(\theta^*(s)^2)\|_{2,\mu}^{1/2}\|\theta(t)^2\|_{2,\mu}^{1/2}$$
$$\leq \|\theta^*(s)^2\|_{2,\mu}^{1/2}\|\theta(t)^2\|_{2,\mu}^{1/2} = \|\theta^*(s)\|_{4,\mu}\|\theta(t)\|_{4,\mu} \leq \|\theta^*\|_{4,\mu}\|\theta\|_{4,\mu},$$

where the fact that Markov semigroups are always contractions on $L^p$ for all $1 \leq p \leq \infty$ has been used. Under the assumptions on $\theta$ and $\theta^*$ we therefore get

(10.7)
$$\mathbf{E}[f(Z_s, Z_t)] \leq C\{\mathbf{E}^X[f^2(X_s, X_t)]\}^{1/2}.$$

It is well known and easily seen that the norm $\| \cdot \|$ in $M$ is in all $L^p(\mu)$, $p < \infty$, and

(10.8)
$$\mathbf{E}^X[\|X_t - X_s\|^{2n}] = C_n|t - s|^n,$$

so
(10.9)
$$\mathbf{E}[\|Z_t - Z_s\|^{2n}] \leq C\{\mathbf{E}^X[\|X_t - X_s\|^{4n}]\}^{1/2} = C\{C_n|t-s|^{2n}\}^{1/2} = c_n|t-s|^n.$$

In view of the Kolmogorov continuous version theorem, this inequality shows that for any $n$, $t \to Z_t$ is a.s. Hölder continuous of any order less than $(n - 1)/(2n)$. $\square$

*(10.10) Remarks.* It is clear that this result can be derived under more general $L^p$-assumptions on $\theta$ and $\theta^*$. Furthermore, if one wants to obtain moment estimates for $Z$, sharper results can be obtained using the hypercontractivity of $p_t$. See [19].

We shall now consider the generators for $Z$, as in Section 5. It is clear that the formula (5.3) holds for $Df$, provided $f$ and $\theta(t)$ have appropriate "regularity." To this end we introduce the Sobolev spaces $W^{m,p}(M,\mu)$ for $m \in \mathbb{N}$ and $1 \leq p < \infty$ with norms

$$(10.11) \qquad \|f\|_{m,p,\mu}^p = \sum_{k=0}^{m} \int_M (\nabla^k f | \nabla^k f)^{p/2} \, d\mu.$$

Here $\nabla$ and $(\cdot|\cdot)$ refers to $\mathcal{H}^{1/2}$ and the term $(\nabla^k f | \nabla^k f)$ to the induced inner product on the $k$-fold tensor product of $\mathcal{H}^{1/2}$.

We define $\mathcal{A} \equiv \bigcap_{p>1} W^{2,p}(M,\mu)$. If $\theta \in \mathcal{A}$, then this holds for $\theta(t)$ too. If also $f \in \mathcal{A}$, then $f\theta(t) \in \mathrm{Dom}\,(H)$, and $Df = -Hf + (\nabla\theta(t)|\nabla f)/\theta(t)$. This is perhaps most clearly seen from formula (5.12), since $H$ can be written $\frac{1}{2}\nabla^*\nabla$, where $\nabla^*$ denotes the divergence given by $\mu$ and the directions of $\mathcal{H}^{1/2}$. The extension to time dependent functions is straight-forward: just consider tensor products of smooth functions on $I$ and functions in $\mathcal{A}$. Then (5.10) results. Similar reasoning shows how to handle $D_*$.

The calculations corresponding to Section 7 apply since we can reduce to finite dimension by considering $\langle Z, \xi \rangle$, for $\xi$ in $\mathcal{S}(\mathbf{R}^d)$. Then $DD\langle Z, \xi \rangle = \langle Z, \Lambda^2 \xi \rangle = \langle Z, (-\Delta + m^2)\xi \rangle$, i.e.

$$(10.12) \qquad DDZ = -\Delta Z + m^2 Z.$$

This is a quantised Euclidean version of the equation for a free classical (scalar) field, in the sense that if the paths $t \to Z_t \equiv \varphi_t$ where smooth, then Eq. (10.12) would reduce to the Euclidean Klein-Gordon equation

$$(10.13) \qquad \frac{\partial^2 \varphi_t}{\partial t^2} + \Delta\varphi_t - m^2\varphi_t = 0.$$

This is another justification, a posteriori, for the regularisations used in EQM (cf. also [5]).

The path integral representation is more problematical. Formally the Lagrangian is $\frac{1}{2}|DZ|^2 + \frac{1}{8}(|\Lambda Z|^2 - \mathrm{Tr}\,\Lambda)$, an expression which is difficult to make sense of. Of course we can obtain a sequence of path integral formulae defining $\theta(t)$ implicitly by using (8.3) together with suitable finite dimensional projections. Already in this case it seems difficult to imagine that (8.3) can be deduced in infinite dimension without renormalisation.

The study of infinite dimensional Bernstein processes is still at its very beginning. A more complete account will appear later in [6].

As should be clear by now, we hope to show that the understanding of quantum physics in the sense initiated by Feynman, can indeed benefit from reflections on its probabilistic foundations. New results, e.g. on the semiclassical limit of EQM, suggest that this hope is well founded [23].

*Acknowledgements.* The first-named author thanks Ana Bela Cruzeiro for the invitation to Lisbon and for friendly and generous hospitality during the conference.

*References.*
1. S. Albeverio and R. Høegh-Krohn: *Dirichlet forms and diffusion processes on rigged Hilbert spaces.*, Z. Warsch.theorie verw. Geb. **40** (1977), 1-57.
2. S. Albeverio, K. Yasue, and J.-C. Zambrini: *Euclidean quantum mechanics: analytical approach.* Ann. Inst. H. Poincare, Physique Théorique **49:3** (1989), 259-308.
3. A. Beurling: *An automorphism of product measures.* Ann. Math. **72** (1960), 189-200.
4. R.M. Blumenthal and R.K. Getoor: "Markov processes and potential theory", Academic Press, New York/London, 1968.
5. E. Carlen: *The stochastic mechanics of free scalar fields.* In A. Truman and I.M. Davies (Eds.), Proc. Swansea Conf. Lect. Notes in Math. 1325, Springer 1988.
6. A.B. Cruzeiro, T. Kolsrud, and J.-C. Zambrini: *Euclidean Quantum Mechanics in infinite dimension*, in preparation.
7. A.B. Cruzeiro and J.-C. Zambrini: *Malliavin calculus and Euclidean quantum mechanics. I. Functional calculus.* J. Funct. Anal., to appear.
8. A.B. Cruzeiro and J.-C. Zambrini: *Feynman's functional calculus and stochastic calculus of variations*, in these proceedings.
9. C. Dellacherie and P.-A. Meyer: "Probabilités et potentiel", vol I, Hermann, Paris, 1975.
10. R. Feynman: Rev. Mod. Phys. **20** (1948).
11. R. Feynman and A.R. Hibbs: "Quantum mechanics and path integrals", McGraw-Hill, New York, 1965.
12. M. Fukushima: "Dirichlet forms and Markov processes", North-Holland and Kodansha, Amsterdam and Tokyo, 1980.
13. N. Ikeda and S. Watanabe: "Stochastic differential equations and diffusion processes", 2nd ed., North-Holland and Kodansha, Amsterdam and Tokyo, 1989.
14. B. Jamison: *Reciprocal processes.* Z. Warsch.theorie verw. Geb. **30** (1974), 65-86.
15. T. Kolsrud: *Gaussian random fields, infinite dimensional Ornstein-Uhlenbeck processes, and symmetric Markov processes*, Acta Appl. Math. **12** (1988), 237-263.
16. E. Nelson: *The free Markov field*, J. Funct. Anal., **12** (1973), 211-227.
17. E. Nelson: "Quantum fluctuations", Princeton Univ. Press, Princeton, 1985.
18. E. Schrödinger: *Sur la théorie relativiste de l'électron et l'interprétation de la méca-nique quantique*, Ann. Inst. H. Poincaré, Physique Théorique **2** (1932), 269-310.

**19**. B. Simon: "The $P(\phi)_2$ Euclidean (quantum) field theory", Princeton Univ. Press, Princeton, 1974.

**20**. J.-C. Zambrini: *Probability and analysis in quantum physics*. In "Stochastic analysis, path integration and dynamics", K.D. Elworthy and J.-C. Zambrini, eds. Pitman Res. Notes in Math. 200, 1989.

**20**. A. Wakolbinger: *A simplified variational characterisation of Schrödinger processes*. J. Math. Phys. **30:12** (1989), 2943-6.

**21**. J.-C. Zambrini: *Probability and analysis in quantum physics*. In "Stochastic analysis, path integration and dynamics", K.D. Elworthy and J.-C. Zambrini, eds. Pitman Res. Notes in Math. 200, 1989.

**22**. J.-C. Zambrini: J. Math. Phys. **27 (4)**, 2307 (1986).

**23**. T. Kolsrud and J.-C. Zambrini: *An introduction to the semiclassical limit of Euclidean Quantum Mechanics*, to appear.

Department of Mathematics,
Royal Institute of Technology,
S-100 44 Stockholm, Sweden
and
CFMC,
Av. Prof. Gama Pinto 2,
1699 Lisboa, Codex, Portugal

# Naturality of Quasi-Invariance of Some Measures

## Paul Malliavin

The purpose of this paper is to show by examples that, in a noncommutative setting, many natural measures do not enjoy the quasi-invariance à la Cameron–Martin which comes immediately to mind. Noncommutativity means the nontriviality of the bracket of the natural vector fields involved. In finite dimension brackets of the natural vector fields associated with an *elliptic operator* do not increase the tangent space, which is, by the ellipticity assumption, generated already by those vector fields. In infinite dimension the phenomena is quite different: starting with an elliptic operator (for instance the Laplacian on a Riemann Hilbert manifold $M$), the bracket phenomena will produce immediately a new tangent space, which could not necessarily contain the given tangent space of $M$. A well-known fact is that measures associated to the Brownian motion on $M$ have to be realized as a Borelian measure on a bigger space than $M$. In the same way, a new tangent space to $M$ has to be realized.

A construction of a new tangent space has been proposed in [5]. The purpose of this paper is to emphasize this construction in some specific examples and to show the failure of the elementary point of view. (See also [1] for some other examples of non quasi-invariance).

## CONTENTS

### 1. Divergence and infinitesimal quasi-invariance, tangent space, integrated quasi-invariance.

Given a measure $\mu$ on a a space $X$ of infinite dimension, a basic concept is the notion of *divergence relativity* to $\mu$ of a vector field $A$. It is denoted by $\operatorname{div}_\mu(A)$ and it is defined by the identity

$$* \qquad \int < A, d\varphi > d\mu = \int \varphi \operatorname{div}_\mu(A) d\mu$$

which has to be valid for a class $\mathcal{C}$ of "smooth" functions $\varphi$. We shall always suppose that $\mathcal{C}$ is dense in $L^2$ and that $\operatorname{div}_\mu(A) \in L^2$. Then the identity * determined without ambiguity the divergence, when it exists.

We shall call tangent bundle along $\mu$ the following data.

(i) At $\mu$ a.e. point $x \in X$ is defined a tangent Hilbert space $T_x(X)$.

(ii) A covariant derivative $\nabla$ is defined on the sections of $T(X) = \cup_x T_x(X)$.

Given $A$ a section of $T(V)$, we shall associate $(\nabla_h A)_x$ which is a bilinear form on $T_x(V)$. We shall use Hilbert Schmidt norms. Then we shall say that $A \in W^{1,p}$ if

$$\int_X \left( \|A_x\|^p_{T_X(V)} + \|(\nabla A)_x\|^p_{T_x(V) \otimes T_x(V)} \right) d\mu(x) < +\infty;$$

then the measure $\mu$ is said to be *regular* if $\mathrm{div}(A)$ exists for all $A \in \bigcap_p W^{1,p}$.

The purpose of this paper is to study two examples which show that the tangent space associated to a measure is essentially unique.

When the measure $\mu$ is regular, the next step is to try its quasi-invariance under a flow generated by a vector field having divergence. There is in this direction the basic theorem of A. B. Cruzeiro [2], which follows:

*Assume that $A$ generates a flow $U_{A,t}$, that $div_\mu(A)$ exists and that*

$$\int \exp\left(c|div_\mu(A)|\right) d\mu < +\infty \quad \text{for all} \quad c;$$

*then $(U_{A,t})_* \mu$ is absolutely continuous relativity to $\mu$, and the Radon Nikodym derivative belongs to all the $L^p$.*

This theorem plays a paramount role in the proof of quasi-invariance of the Wiener measure on loop group [3]. It is also possible to find in [2] a sufficient condition under which the $U_{A,t}$ flow exists.

## 2. Path space over a non compact Lie group

Given a connected Lie group $G$ of matrices, we denote $\mathfrak{g}$ its Lie algebra. We choose a Euclidean metric on $\mathfrak{g}$ and we denote by $e_1 \ldots e_n$ an orthonormal basis of $\mathfrak{g}$. Furthermore, we define, for every $\varphi$ smooth function on $G$

$$(\partial_k \varphi)(g) = \left\{ \frac{d}{d\varepsilon} \varphi(g\exp(\varepsilon e_k)) \right\}_{\varepsilon = 0} \qquad \Delta = \frac{1}{2} \sum_k \partial_k^2.$$

We denote by $\mathbb{P}_e(G)$ the space of continuous maps of $[0,1]$ into $G$, starting at the unit $e$ of $G$. The Brownian motion generated by $\Delta$ defines a Wiener measure $\mu_{\mathbb{P}_e}(G)$.

We denote by $\mathbb{P}_0^1(\mathfrak{g})$ the Hilbert space of paths on $\mathfrak{g}$, with finite energy, that is such that

$$\|u\|^2_{\mathbb{P}_0^1(\mathfrak{g})} = \int_0^1 \|\dot{u}(\tau)\|^2_{\mathfrak{g}} d\tau < +\infty.$$

Then we can identify $\mathbb{P}_0^1(\mathfrak{g})$ to the tangent space at the point $\gamma \in \mathbb{P}_e(G)$ in two ways either on the right or on the left.

For the left action we have proved [3], [4].

**Theorem.** *There exists $K_u \in \bigcap_{p<+\infty} L^p$ such that for every regular cylindric function $\psi$ we have*

$$\frac{d}{d\varepsilon}\left\{ \int \psi(\exp(\varepsilon u)\gamma) d\mu_{\mathbb{P}_e(G)}(\gamma) \right\}_{\varepsilon=0} = \int \psi K_u d\mu_{\mathbb{P}_e(G)}.$$

We want to prove the following negative result.

**Theorem.** *Assume that the Adjoint representation of $G$ is not unitary, then the measure $\mu_{\mathbb{P}_e(G)}$ is not quasi invariant under the right infinitesimal action of $\mathbb{P}_0^1(\mathfrak{g})$.*

**Proof.** We have to prove that given $u \in \mathbb{P}_0(\mathfrak{g})$, it does not exist $K \in L^1$ such that

$$\left\{ \frac{d}{d\varepsilon} \int \psi(\gamma \exp(\varepsilon u)) d\mu_{\mathbb{P}_e(G)} \right\}_{\varepsilon=0} = \int \psi K d\mu_{\mathbb{P}_e(G)}.$$

We denote by $X$ the Wiener space defined by the Brownian motion on $\mathfrak{g}$. Then as vector space $X = \mathbb{P}_0(\mathfrak{g})$ and the Wiener measure is $\mu_{\mathbb{P}_0}(\mathfrak{g})$ that we shall also shorthanded as $\mu_x$ or $\mu$. Then the Itô map $\Phi : X \to \mathbb{P}_e(G)$ is defined by solving the following SDE

$$\begin{cases} dg_x & = g_x[\Sigma e\, dx^k + c\, dt] \\ g_x(0) & = e \end{cases}$$

where $c = \frac{1}{2}\Sigma e_k^2$.

We denote $\Phi(x)(\cdot) = g_x(\cdot)$. Then $\Phi$ is a Borelian bijection sending the measure $\mu$ to $\mu_{\mathbb{P}_e(G)}$. Therefore the quasi invariance problem for $\mu_{\mathbb{P}_e(G)}$ will be transferred by $\Phi^{-1}$ to a similar problem for $\mu$.

Given $h \in H$=Cameron–Martin space=$\mathbb{P}_0^1(\mathfrak{g})$, we denote

$$\left\{ \frac{d}{d\varepsilon}\Phi(x+\varepsilon h) \right\}_{\varepsilon=0} \Phi^{-1}(x) = k$$

$$(\Phi(x))^{-1}\left\{ \frac{d}{d\varepsilon}\Phi(x+\varepsilon h) \right\}_{\varepsilon=0} = \ell.$$

The computation of $k$ depends upon the following familiar procedure of linearization. Define $g_{x,\varepsilon}$ as the solution of the following SDE

$$\begin{cases} dg_{x,\varepsilon} & = g_{x,\varepsilon_k}(\Sigma e_k\, dx^k + (c + \varepsilon \dot{h})d\tau) \\ g_{x,\varepsilon}(0) & = e \end{cases}$$

Differentiating in $\varepsilon$ and denoting $M = \left\{ \frac{\partial g_{x,\varepsilon}}{\partial \varepsilon} \right\}_{\varepsilon=0}$, we get

$$\begin{cases} dM &= g_x \dot{h} d\tau + M(\Sigma e_k dx^k + c d\tau). \\ M(0) &=0 \end{cases}$$

We have

$$dg_x = (-\Sigma e_k dx^k + c d\tau) g_x^{-1}.$$

Therefore by Itô calculus

$$dk(\tau) = g_x(\tau)\dot{h}(\tau)g_x^{-1}(\tau)d\tau.$$

We have the remarkable fact that the stochastic differential equation that we would in principle obtain for $k$ degenerates into an ordinary differential equation.

We shall denote $a^g = gag^{-1}$ and the equation for $k$ can be read therefore

$$\dot{k} = (\dot{h})^{g_x} \quad \text{or} \quad \dot{h}(\dot{k})^{g_x^{-1}}.$$

The parallel computation for $\ell$ will lead to a *true stochastic differential equation*. Instead of computing this equation we shall write its integrated form. We have

$$\ell = k^{g_x^{-1}}.$$

As

$$k(\tau) = \int_0^\tau \dot{k}(\xi)d\xi = \int_0^\tau (\dot{h}(\xi))^{g_x(\xi)}d\xi,$$

we get

$$\ell(\tau) = \int_0^\tau [\dot{h}(\xi)]^{g_x(\xi)g_x^{-1}(\tau)}d\tau.$$

Now the first theorem is proved as in [4] taking

$$k_u(g_x) \equiv \int_0^1 ((\dot{u}(\tau))^{g_x^{-1}(\tau)}|dx(\tau))_\mathfrak{g}.$$

Now by the transport of structure the right infinitesimal quasi-invariance by $u$ is transformed in the quasi-invariance of the Wiener measure by the following vector field $Z$ defined by

$$\dot{Z}_x(\tau) = \left( \frac{d}{d\tau} \left[ u(\tau)^{g_x(\tau)} \right] \right)^{g_x^{-1}(\tau)}.$$

Even for $u$ real analytic the derivatives at the r. h. s. do not exist. This impossibility to apply the standard Cameron–Martin machinery does not disprove of course the quasi-invariance. Furthermore, when $G$ is compact, we have quasi-invariance on the left or on the right, granted the symmetry of the measure $\mu_{\mathbb{P}_e(G)}$. Nevertheless the previous formula has, even in this case, no meaning. To disprove quasi-invariance we will have to use the following procedure of discretization.

We denote by $S_n$ the set $\{k2^{-n}\}$, $k = 1, \ldots, 2^n$, by $\mathcal{E}_n$ the evaluation map which send $P_e(G)$ to $G^{S_n}$. Denote $(\mathcal{E}_n)_* \mu_{\mathbb{P}_e}(G)$ by $\mu_n$. Then

$$d\mu_n = \overset{2^n-1}{\underset{i=0}{\otimes}} p_{2^{-n}}(g_i^{-1} g_{i+1}) dg_{i+1}$$

where $dg$ is a right invariant Haar measure.

Denote by $\ell_n$ the restriction of $\ell$ to $S_n$. Then $\ell_n \in \mathfrak{g}^{S_n}$ and we have

$$\operatorname{div}_{\mu_n}(\ell_n) = \sum_{j=0}^{2^n-1} B_n^j$$

with $B_j^n$ is defined by

$$B_n^j = u_n(\ell(j+1)2^{-n}) - Ad(g_j^{-1}g_{j+1})\ell(j2^{-n}), g_j^{-1}g_{j+1})$$

with

$$u_n(z, g) = \left\{ \frac{d}{d\varepsilon} \log(p_{2^{-n}}(g \exp(\varepsilon z))) \right\}_{\varepsilon=0}.$$

Now if $\mu_{\mathbb{P}_e}(G)$ will be quasi-invariant under the right action of $\ell$, this will imply that $Y_n = \{\operatorname{div}_{\mu_n}(\ell_n)\}$ will be an $L^1$-martingale. The convergence of this $L^1$-martingale is equivalent to

(i) the almost sure convergence of $\Sigma[Y_{n+1} - Y_n]^2$.

We shall disprove this statement by the two following facts

(ii)    $Z_n = (Y_{n+1} - Y_n)$ are independent random variables.
(iii)    $E(Z_n^2) \sim (\int_0^1 |\ell(\tau)|^2 d\tau)2^n$.

In fact, statements (ii) and (iii) will be proved for a new sequence $\tilde{Z}_n$ which will verify $\Sigma|\tilde{Z}_n - Z_n|^2 < +\infty$ almost surely.

The statement (i) for $\tilde{Z}_n$ will propagate to $Z_n$ according the inequality

$$Z_n^2 = (\tilde{Z}_n + Z_n - \tilde{Z}_n)^2 > [\tilde{Z}_n^2 - (\tilde{Z} - Z_n)^2].$$

In order to construct $\tilde{Z}_n$ we remark that

$$u_n(z, g) \sim -2^n(z|\log g) \text{ when } n \to +\infty.$$

We have $Y_n = \Sigma D_n^j$ with

$$D_n^j = u_n(\ell(j2^{-n}) - Ad(g_j^{-1}g_j)\ell(j+1)2^{-n}), g_j^{-1}g_{j+1}).$$

We remark that

(iv)    $2^n \sum_j [\ell(j2^{-n}) - \ell((j+1)2^{-n})]^2 \longrightarrow \int_0^1 [\dot{\ell}(\tau)]^2 d\tau < +\infty.$

Then according (iv) it is allowed to replace in the expression of $D_n^i \ell((j+1)2^{-n})$ by $\ell(j2^{-n})$.

For $\tau$ small $\log(g_x(\tau))$ behave as the Brownian motion on $\mathfrak{g}$. We shall use the Paul Levy construction of the Brownian motion. The innovation to go from the data of the Brownian motion on $\{j2^{-n}\}$ to the data to $\{q2^{-n-1}\}$ is the data for each interval of a Gaussian variable $I_j$ of variance $2^{-n}$. We define

$$\tilde{Z}_n = \sum_j C_n^j \quad \text{with} \quad C_n^j = 2^n (ad(I_j)\ell(j2^{-n})|\Gamma_j)_{\mathfrak{g}^*}.$$

We choose an orthonormal basis $e_1 \ldots e_n$ of $\mathfrak{g}$, then denote by $c_{i,k,\ell}$ the constants of structure of $\mathfrak{g}$ in this basis. Denote by $\Gamma_j^k$ the composant of $\Gamma_j$ in this base, we have

$$C_n^j = 2^n \Sigma c_{i,k,\ell} \ell^k (j2^{-n}) \Gamma_j^\ell \Gamma_\ell^i$$
$$= 2^n \sum_{i<\ell} (c_{i,k,\ell} - c_{\ell,k,i}) \ell^k (j2^{-n}) \Gamma_j^\ell \Gamma_j^i.$$

Define

$$\gamma_{k,k'} = \sum_{i,\ell} (c_{i,k,\ell} - c_{\ell,k,i})(c_{i,k',\ell} - c_{\ell,k',i}),$$

then we have

$$E((c_n^j)^2) = \Sigma \gamma_{k,k'} \ell^k (j2^{-n}) \ell^{k'} (j2^{-n}).$$

We have proved (iii) or the following more specific form

(v)    $E(\tilde{Z}_n^2) \sim 2^n \int_0^1 \gamma_{k,k'} \ell^k(\tau) \ell^{k'}(\tau) d\tau,$

then if the positive matrix $\gamma_{k,k'}$ is not zero we shall choose $\ell(\tau)$ in the orthogonal of its kernel for all $\tau > 0$ and we shall get the proof.

Now, if $\gamma_{k,k'} = 0$, this means that the adjoint action of $G$ under $\mathfrak{g}$ is orthogonal. Then in this case the argument of [3], Proposition 2.2.5 can be carried over, the measure is symmetric, the left quasi-invariance under the action of $\mathbb{P}_0^1(\mathfrak{g})$ implies the right invariance.

## 3. Paths on a Riemannian manifold.

The reader could object about the construction of the last paragraph that quasi-invariance on the left is the natural concept and that quasi-invariance on the right is a nasty point of view which is not natural.

On a non homogeneous Riemannian manifold $V$, there is only one possible definition for a natural infinitesimal deformation of a path.

### 3.1. *Vector field on loop of finite energy.*

We recall that the covariant derivative $\nabla$ defines along smooth curves an absolute derivative. Therefore it is possible to define the $H^1$-variation $u$ along a $C^\infty$ curve $\varphi(s)$ as a section of the tangent space above $\varphi(s)$ satisfying

(i) $\qquad \|u\|_{H^1}^2 = \int \|\nabla u\|^2(\varphi(s))ds$.

In local coordinate $v^i$ the variation is written by its component

$$u = \Sigma u^i \frac{\partial}{\partial v^i}.$$

Then the component of $\nabla u$ is expressed through Christofell symbols:

(ii) $\qquad (\nabla u)^i = \frac{d}{ds}u^i(\varphi(s)) + \Gamma_{j,k}^i(\varphi(s))\dot{\varphi}^k(s)u^i$,

then we have $\|\nabla u\|^2 = g_{ij}(\nabla u)^i(\nabla u)^j$ and the formula (i) has a proper meaning.

We denote by $\mathbb{P}_{v_0}^1$ the paths on $V$ starting at $v_0$ with finite energy that is such that

$$\int_0^1 \|\dot{\varphi}(s)\|^2 ds < +\infty,$$

then the tangent space at the point $\varphi \in \mathbb{P}_{v_0}^1(V)$ is the variation satisfying (i) and (ii) which have still a perfect meaning. We have to add the condition $u(0) = 0$. We have to define the tangent space

$$T(\mathbb{P}_{v_0}^1(V)) = \bigcup_{\varphi \in \mathbb{P}^1(V)} T_\varphi(\mathbb{P}^1(V)).$$

We want now to define a connection on $T(\mathbb{P}^1(V)))$. (See also [5]). Given a curve on $\mathbb{P}^1(V))$ we want to define the covariant derivative along this curve.

A curve on $\mathbb{P}^1(V)$ is in fact a map $\Phi[0,1]^2 \to V$; then, for $t$ fixed, the partial map $\Phi_t(s)$ defines a curve.

In the same way, a vector field along the curve $\Phi_t$ is defined as for every fixed $t$ a vector field along $\Phi_t$. We shall write this notation in a local chart. Then we have two $\mathbb{R}^n$-valued function $\Phi(t,s)$, $u(t,s)$, the two partial functions $\Phi_t$, $u_t$ denoting a point of $\mathbb{P}^1(V)$ and the tangent vector at this point.

The map $(t,s) \to \Phi(t,s)$ defines a "surface" in $V$. The covariant derivative, acting on $T(V)$, can be taken along this surface. We shall denote by $\nabla_s, \nabla_t$ the covariant derivative along the curves $t$ =constant, $s$ =constant.

As $u(t,0) = 0$, we have the fact that $\nabla_s u$ determines $u(t,0)$. The variation of $u$ along $t$ can be therefore defined as the derivative $\nabla_t(\nabla_s u)$.

We define the covariant derivative on $u_t$ by

$$(\overline{\nabla}u_t)(s) = \int_0^s (\nabla_t \nabla_s u)(t,\xi)d\xi,$$

we have

(iii)     $\frac{d}{dt}(\int_0^1 \|\nabla_s u\|^2 ds) = 2\int_0^1 (\nabla_t \nabla_s u | \nabla_s u) ds.$

then according to (iii) the length is preserved.

## 3.2. Brownian paths.

We want to develop for the Brownian the preceding section. We will formulate the preceding construction in terms of ordinary differential equation. Then the *transfert principle* [5] will give their extension by Stratanovitch stochastic differential equation. We want to realize this program in an intrinsic setting.

We denote by $0(V)$ the orthonormal frame bundle over $V$, by $A_1 \ldots A_n$ the canonic horizontal vector fields on $0(V)$ defined by Riemannian connection. Given $x \in X$. We associate $r_x \in \mathbb{P}_{r_0}(0(V))$ defined by the following SDE

$$\begin{cases} dr_x & = r_x o \sum_{k=1}^n A_k dx^k \qquad \text{(Stratanovitch SDE)} \\ r_x(0) & = r_0 \end{cases}$$

where $r_0$ is a fixed frame of $T_{v_0}(V)$ (this means a isometry of $R^n$ on $T_{v_0}(V)$) and where $x(\tau)$ is the Brownian motion on $\mathbb{R}^n$.

We denote by $p : 0(V) \to V$ then $p(r_x(\tau))$ is the Brownian motion on $V$. We denote by $\Phi$ the Itô map $x \to p \circ r_x$:

$$\Phi : X \to \mathbb{P}_{v_0}(V).$$

The motion $\tau \to r_x(\tau)$ describes the motion of a frame in the parallel transport along $v_x(\tau)$. Therefore we have

**Proposition 3.2.1.** *The tangent vectors at the point $v_x$ are of the form:*

$$u(\tau) = r_x(\tau)v(\tau) \quad \text{with} \quad v \in H.$$

**Theorem 3.2.2.** *The Itô map $\Phi$ is bijective, it does not preserve the tangent spaces.*

**Proof.** The tangent space at $X$ is the Cameron–Martin space $H = H^1([0,1]; R^n)$. Given $h \in H$ we have to solve the following SDE in order to compute $\Phi'(x)$:

$$dr_{x+\varepsilon h} = r_{x+\varepsilon h} \circ (A_k(dx^k + \varepsilon h^k d\tau)).$$

In order to compute derivatives we shall use the natural parallelism on $0(V)$. To the horizontal vector field are associated the horizontal 1-differential forms $\pi^i$ defined on $0(V)$.

Furthermore the orthogonal group operates on the frames on the right: $r \to r \circ g$. The tangent space to $p^{-1}(v_0)$ can therefore be defined to the Lie algebra of $0(n)$. This identification leads to the vertical differential from $\pi_j^i$ defined on $0(V)$, (antisymmetric in the indices $i, j$); then the structural equations of Riemannian geometry are

(i) $\quad - < z_1 \wedge z_2, d\pi^i > = \pi_j^i(z_1)\pi^j(z_2) - \pi_j^i(z_2)\pi^j(z_1)$

$\quad < z_1 \wedge z_2, d\pi_j^i > + \pi_q^i(z_1)\pi_j^q(z_2) - \pi_q^i(z_2)\pi_j^q(z_2)$

$\quad = R_{j,k,\ell}^i \pi^k(z_1)\pi^\ell(z_2)$

where $R_{j,k,\ell}^i$ is the curvature tensor on $V$.

We fix a mollifier $\chi$, that is a smooth function of integral 1 with support contained in $[0,1]$. We denote $\chi_\eta(\tau) = \int x(\tau - \eta\lambda)\chi(\lambda)d\lambda$.

Now the stochastic moving frame $r_x$ is approached by the equation:

$$< dr_{x_\eta + \varepsilon h}, \pi^i > = dx_\eta^i + \varepsilon h d\tau \quad < dr_{x + \varepsilon h}, \pi_j^i > = 0.$$

Consider the sheet

$$\psi_{x,\eta} : (\tau, \varepsilon) \mapsto r_{x_\eta + \varepsilon h}(\tau).$$

We denote $\theta_\eta = (\psi_x)^* \pi$, then

$$\theta_\eta = a_\eta(\tau, \varepsilon)d\tau + b_\eta(\tau, \varepsilon)d\varepsilon$$

$$a_\eta(t, \varepsilon)d\tau = dx_\eta + \varepsilon h d\tau$$

$$\frac{\partial a_\eta}{\partial \varepsilon}d\tau = \dot{h}d\tau;$$

therefore

$$d\theta_\eta = \left[ \dot{h} - \frac{\partial b_\eta}{\partial \tau}s \right] d\varepsilon \wedge d\tau;$$

from the other hand

$$d\theta_\eta = (\psi_{x,\eta})^* d\pi$$

and $d\pi$ can be expressed through the structural equation (i). Then if we denote $k_\eta(\tau) = b_\eta(\tau, 0)$, we get

$$\dot{h} - \dot{k}_\eta = < \dot{x}_\eta \wedge k_\eta, d\pi >.$$

Denote $i$ the interior product, we have

$$\dot{k}_\eta + < i(\dot{x}_\eta)d\pi, k_\eta > = \dot{h}.$$

This equation is solved by the Lagrange method of variation of constants. Consider the matrix valued differential equation

(ii) $\quad \frac{d}{d\tau}J_{x,\eta}(\tau, \tau_0) = (i(\dot{x}_\eta)d\pi) \cdot J_{x,\eta}(\tau, \tau_0)$

$\quad J_{x,\eta}(\tau_0, \tau_0) = $ Identity;

then

(iii)    $k_\eta(\tau) = \int_0^\tau J_{x,\eta}(\tau, \tau_0) \dot{h}(\tau_0) d\tau_0$.

It is possible to express $J$ in terms of the Riemannian curvature. Denote $D(n)$, the Lie algebra of the Euclidean motion in $\mathbb{R}^n$

$$D(n) = R^n \oplus M_{n,a}$$

where $M_{n,a}$ are the $n \times n$ antisymmetric matrices.

Define $u_j \in End(D(n))$ by

$$u_j(z^i) = z^i_j \quad (u_j(z))^i_q = R^i_{q,j,\ell} z^\ell;$$

then (i) implies that (ii) can be written as (see [5]):

(iv)    $\frac{d}{d\tau} J_{x,\eta} = (\Sigma u_j \dot{x}^j_\eta) J_{x,\eta}$.

As $\eta \to 0$ the equation (iv) tends to the corresponding SDE in the Stratanovitch sense and therefore (iv) and (iii) show the following lemma (a proof going in the detail can be found in [5]).

**Lemma 3.2.3.** *When* $\eta \to 0$, $k_\eta$ *converges in* $H$ *to* $k$.

We have now:

**Theorem 3.2.4.** *The natural variation of a path does not induce a Cameron–Martin vector on the Wiener space.*

**Proof.** Given $q \in H(\mathbb{R}^n)$ the vector field associated to $q$ at $v_x(\cdot)$ is given by

$$z(\ ) = p'(\sum_k A_k q^k(\cdot)).$$

We want to find $h \in H^1$ such that

$$\int_0^\tau J(\tau, \xi) \dot{h}(\xi) d\xi = q(\tau).$$

This equation is *impossible*: the l.h.s. being only an $H^{1/2-\varepsilon}$ when the r. h. s. is in $H^1$. This proves the theorem.

## Conclusion

We have two possible variations of a path of the Brownian motion on a Riemannian manifold.

(i) The variation defined by the *development* that is by the variation of the Itô stochastic differential equation; then this variation leaves the Wiener measure quasi invariant. It has the inconvenience to depend upon the *starting point*. In the case of $\mathbb{L}(V)$ the space of *free loops* (that is the space of continuous map of the circle into $V$), it is quite clear that the choice of a starting point would be quite unnatural.

(ii) The other definition of the tangent space is the sections along $v_x(\ )$ of the tangent bundle which have their *covariant* derivatives in $L^2$. This definition is quite natural. The problem of quasi invariance cannot be reduced to the Cameron Martin Theorem through the Itô map. A more elaborate approach seems needed [7].

## REFERENCES

[1] Boragachev A., *Comptes rendus*, tome 311, p. 807, décembre 1990.

[2] Cruzeiro A. B., *Equations différentielles sur l'espace de Wiener et formules de Cameron-Martin non linéaires*, J. Funct. Anal., **54**(1983), p. 206–227.

[3] Malliavin M.-P. and Malliavin P., *Integration on loop groups*, J. Funct. Anal., **93**(1990), p. 208–236.

[4] Malliavin M.-P. and Malliavin P., *Mesures quasi invariantes sur des groupes de dimension infinite*, Comptes rendus, tome 311, p. 765, décembre 1990.

[5] Malliavin P., *Diffusion on the loops*, Conference in honor of A. Zygmund, Chicago, 1983 (Editors W. Beckner, A. P. Calderon, R. Fefferman, Peter Jones, p. 764–778.

[6] Malliavin P., *Hypoellipticity in infinite dimension*, Proceeding, Conference North Western, november 1989 (M. Pinsky editor), Birkhaüser, 1990.

[7] Malliavin P., *Intrinsic stochastic calculus on a Riemannian manifold*, To appear.

Paul Malliavin
10 rue Saint-Louis en l'Isle
75004 PARIS

# STOCHASTIC DIFFERENTIAL EQUATIONS
# WITH BOUNDARY CONDITIONS

*by*

David Nualart
Facultat de Matemàtiques
Universitat de Barcelona
Gran Via, 585
08007 Barcelona - Spain.

Etienne Pardoux
Mathématiques, URA 225
Université de Provence
13331 Marseille Cedex 3
France

## 1. Introduction

The stochastic calculus with anticipating integrands has been recently developed by several authors (see in particular [5,6,9] and the references therein). This new theory allows to study different types of stochastic differential equations driven by a $d$-dimensional Brownian motion $\{W(t), 0 \leq t \leq 1\}$, where the solutions turn out to be non necessarily adapted to the filtration generated by $W$. We refer the reader to [12] for a survey of the applications of the anticipating stochastic calculus to stochastic differential equations. In particular one can consider stochastic differential equations of the form

$$dX_t = f(X_t) + \sum_{i=1}^{k} g_i(X_t) \circ dW_t^i, \qquad 0 \leq t \leq 1, \tag{1.1}$$

and, instead of giving the value of the process at time zero, we impose a boundary condition of the form $h(X_0, X_1) = h_0$. In general, the solution $\{X_t, 0 \leq t \leq 1\}$ will not be an adapted process, and the stochastic integral $\int_0^t g_i(X_s) \circ dW_s^i$ is taken in the extended Stratonovich sense. The existence and uniqueness of a solution for an equation of this type has been investigated in some particular cases, and the Markov property of the solution has been studied. More precisely the following particular situations have been considered:

(a) The functions $f$, $g$ and $h$ are affine. (Ocone–Pardoux [11])

(b) $k = d$ and the function $g$ is a constant equal to the identity matrix. (Nualart–Pardoux [7])

(c) $k = d = 1$, and $h$ is linear. (Donati–Martin [2])

(d) Second order stochastic differential equations in dimension one of the following type

$$\ddot{X}_t + f(X_t, \dot{X}_t) = \dot{W}_t, \qquad 0 \leq t \leq 1, \tag{1.2}$$

with Dirichlet boundary conditions $X_0 = a$, $X_1 = b$. (Nualart–Pardoux [8]).

The objective of this paper is to describe some of the results obtained in these articles concerning the existence and uniqueness and the Markov property of the solution of the equations (1.1) and (1.2). In order to illustrate the techniques used in the cases (b), (c) and (d) we are going to present a complete study in the particular case of a second order stochastic differential equation of the type (1.2) when the function $f$ depends only on the variable $X_t$. This hypothesis allows to simplify the arguments used in [8], and more direct computations can be carried out in this case.

The organization of the paper is as follows. In Section 2 we will introduce the basic notations and results of the theory of noncausal stochastic calculus that will be needed later. In particular we will state an extended version of the Girsanov theorem for non necessarily adapted processes which is due to Ramer [13] and Kusuoka [3]. In Section 3 we will study the Markov property of the solution of the second order stochastic differential equation

$$\ddot{X}_t + f(X_t, t) = \dot{W}_t, \qquad 0 \le t \le 1,$$
$$X_0 = X_1 = 0.$$
(1.3)

The main result is as follows. Assuming that $f$ has linear growth, admits two continuous partial derivatives with respect to $x$ and $f'_x \le 0$, then the process $\{(X_t, \dot{X}_t)\}$ is Markovian if and only if $f$ is an affine function. The proof of this result is based on the following idea. Denote by $\{Y_t, 0 \le t \le 1\}$ the solution of (1.3) for $f \equiv 0$. Then $\{(Y_t, \dot{Y}_t)\}$ is known to be a Markov process. On the other hand, the law of $\{X_t\}$ is the same as the law of $\{Y_t\}$ under a new probability measure $Q$ which is absolutely continuous with respect to the Wiener measure. Then one can show that for any $t \in (0,1)$ the corresponding Radon–Nikodym derivative cannot be expressed as the product of two factors one of them $\sigma\{(Y_s, \dot{Y}_s), 0 \le s \le t\}$–measurable and the other $\sigma\{(Y_s, \dot{Y}_s), t \le s \le 1\}$–measurable. This lack of factorization prevents for the Markov property to hold.

Section 4 will be devoted to survey some of the results obtained in [2,7,11] on the existence and uniqueness of a solution and its Markov property for different types of first order stochastic differential equations with boundary conditions.

## 2. Some elements of noncausal stochastic calculus.

Let $\Omega = C_0([0,1])$ be the space of continuous functions on $[0,1]$ which vanish at zero. We denote by $P$ the Wiener measure on $\Omega$, and by $\mathcal{F}$ the Borel $\sigma$–field completed with respect to $P$. Then, $W_t(\omega) = \omega_t$, $0 \le t \le 1$ will represent the Wiener process defined on the canonical space $(\Omega, \mathcal{F}, P)$. The Hilbert space $L^2(0,1)$ will be denoted by $H$. Let us first recall briefly the notions of derivation on Wiener space and of Skorohod integral. We denote by $S$ the subset of $L^2(\Omega)$ consisting of those random variables of the form:

$$F = f\left( \int_0^1 h_1(t)dW_t, \ldots, \int_0^1 h_n(t)dW_t \right)$$
(2.1)

where $n \geq 1$, $h_1, \ldots, h_n \in L^2(0,1)$, and $f \in C_b^\infty(\mathbb{R}^n)$. The random variables of the form (2.1) are called *smooth functionals*. For a smooth functional $F \in S$ of the form (2.1) we define its derivative $DF$ as the stochastic process $\{D_t F, 0 \leq t \leq 1\}$ given by

$$D_t F = \sum_{i=1}^n \frac{\partial f}{\partial x_i} \left( \int_0^1 h_1(s)dW_s, \ldots, \int_0^1 h_n(s)dW_s \right) h_i(t). \qquad (2.2)$$

Then $D$ is a closable unbounded operator from $L^2(\Omega)$ into $L^2(\Omega \times [0,1])$. We will denote by $\mathbb{D}^{1,2}$ the completion of $S$ with respect to the norm $\| \cdot \|_{1,2}$ defined by

$$\|F\|_{1,2} = \|F\|_2 + \|DF\|_{L^2(\Omega \times [0,1])}, \qquad F \in S.$$

Similarly, for any fixed $h \in L^2(0,1)$ we can define $D^h F = \int_0^1 D_t F h(t)dt$, if $F \in S$. Then $D^h$ is closable and we will denote by $\mathbb{D}^{h,2}$ the closure of $S$ by the norm

$$\|F\|_{h,2} = \|F\|_2 + \|D^h F\|_2.$$

We will denote by $\delta$ the adjoint of the derivation operator $D$. That means, $\delta$ is a closed and unbounded operator from $L^2(\Omega \times [0,1])$ into $L^2(\Omega)$ defined as follows: The domain of $\delta$, Dom $\delta$, is the set of processes $u \in L^2(\Omega \times [0,1])$ such that there exists a positive constant $c_u$ verifying

$$\left| E \int_0^1 D_t F u_t dt \right| \leq c_u \|F\|_2, \qquad (2.3)$$

for all $F \in S$. If $u$ belongs to the domain of $\delta$ then $\delta(u)$ is the square integrable random variable determined by the duality relation

$$E \left( \int_0^1 D_t F u_t dt \right) = E(\delta(u)F), \qquad F \in \mathbb{D}^{1,2}. \qquad (2.4)$$

The operator $\delta$ is an extension of the Itô integral in the sense that the class $L_a^2$ of processes $u$ in $L^2(\Omega \times [0,1])$ which are adapted to the Brownian filtration is included in Dom $\delta$ and $\delta(u)$ is equal to the Itô integral if $u \in L_a^2$. The operator $\delta$ is called the Skorohod stochastic integral. We refer the reader to [6] for the proof of the basic facts of the stochastic calculus for the Skorohod integral.

Define $\mathbb{L}^{1,2} = L^2([0,1]; \mathbb{D}^{1,2})$. Then the space $\mathbb{L}^{1,2}$ is included into the domain of $\delta$. The operators $D$, $D^h$ and $\delta$ are *local* in the following sense:

(a) $1_{\{F=0\}} DF = 0$, for all $F \in \mathbb{D}^{1,2}$,

(b) $1_{\{F=0\}} D^h F = 0$, for all $F \in \mathbb{D}^{h,2}$,

(c) $1_{\{\int_0^1 u_t^2 dt = 0\}} \delta(u) = 0$, for all $u \in \mathbb{L}^{1,2}$.

Then one can define the spaces $\mathbb{D}_{loc}^{1,2}$, $\mathbb{D}_{loc}^{h,2}$ and $\mathbb{L}_{loc}^{1,2}$ by a standard localization method. For instance, $\mathbb{L}_{loc}^{1,2}$ is the space of processes $u$ such that there exists a sequence $\{(\Omega_n, u_n), n \geq 1\}$ such that $\Omega_n \in \mathcal{F}$, $\Omega_n \uparrow \Omega$, a.s., $u_n \in \mathbb{L}^{1,2}$, and $u_n = u$ on $\Omega_n$ for each $n$. By property (c) the Skorohod integral can be extended to the processes of the class $\mathbb{IL}_{loc}^{1,2}$. In a similar way the operators $D$ and $D^h$ can be properly defined on the sets $\mathbb{D}_{loc}^{1,2}$ and $\mathbb{D}_{loc}^{h,2}$, respectively.

The derivation operator $D$ can be used to translate some measurability properties into algebraic conditions. The following lemma is an example of this application of the derivation operator, and it will be used in the next section.

**Lemma 2.1.** Fix $k$ functions $h_1, \ldots, h_k \in H$, where we recall that $H$ denotes the Hilbert space $L^2(0,1)$. Let $\mathcal{G}$ be the $\sigma$-algebra generated by the Gaussian random variables $\int_0^1 h_1(t) dW_t, \ldots, \int_0^1 h_k(t) dW_t$. Let $F$ be a random variable in the space $\mathbb{D}_{loc}^{1,2}$ such that $F \, 1_G$ is $\mathcal{G}$-measurable for some set $G \in \mathcal{G}$. Then there exist random variables $A_i$, $1 \leq i \leq k$ such that

$$1_G D_t F = 1_G \sum_{i=1}^k A_i h_i(t),$$

for $dP \times dt$ almost all $(\omega, t) \in \Omega \times [0,1]$.

*Proof:* Consider the subspace $K$ of $H = L^2(0,1)$ spanned by $h_1, \ldots, h_k$. Since we can approximate $F$ by $\varphi_n(F)$ with $\varphi_n \in C_b^\infty(\mathbb{R})$ and $\varphi_n(x) = x$ for $|x| \leq n$, it is sufficient to prove the result for $F \in \mathbb{D}_{loc}^{1,2} \bigcap L^2(\Omega)$. Let $h \perp K$. Clearly the conditional expectation $E(F|\mathcal{G})$ belongs to $\mathbb{D}^{h,2}$ and $D^h E(F|\mathcal{G}) = 0$. Moreover the hypotheses of the lemma imply that $F \in \mathbb{D}_{loc}^{h,2}$ and $E(F|\mathcal{G}) = F$ a.s. on $G$. From the local property of the operator $D^h$ we deduce that

$$D^h F = \int_0^1 h_t D_t F \, dt = 0, \qquad \text{a.s.} \quad \text{on} \quad G.$$

Then it only remains to choose a countable dense set in the orthogonal of $K$ and we obtain that $1_G DF \in K$ a.s., which implies the desired result.                    Q.E.D.

Let us recall the definition of the Stratonovich integral and its relation with the Skorohod integral. Let $\{u_t, 0 \leq t \leq 1\}$ be a measurable process such that $\int_0^1 u_t^2 dt < \infty$ a.s. Then $u$ is said to be *Stratonovich integrable* if

$$\sum_{j=1}^{n-1} \left( \frac{1}{t_{j+1} - t_j} \int_{t_j}^{t_{j+1}} u_s \, ds \right) (W(t_{j+1}) - W(t_j))$$

converges in probability as $|\pi| \to 0$, where $\pi = \{0 = t_1 < \cdots < t_n = 1\}$ runs over all finite partitions of $[0,1]$ and $|\pi| = \max_j(t_{j+1} - t_j)$. The limit will be called the *Stratonovich integral* of $u$ and it will be denoted by $\int_0^1 u_t \circ dW_t$.

Let $\mathbb{IL}_C^{1,2}$ denote the set of processes $u \in \mathbb{IL}^{1,2}$ such that:

(i) The set of functions $\{s \to D_t u_s, 0 \leq s \leq t\}$, $t \in [0,1]$, with values in $L^2(\Omega)$ is equicontinuous for some version of $Du$, and similarly, the set of functions $\{s \to D_t u_s, t \leq s \leq 1\}$, $t \in [0,1]$, is also equicontinuous for a version (possibly different) of $Du$.

(ii) ess $\sup_{s,t} E(|D_s u_t|^2) < \infty$.

For a process $u$ in the class $\mathbb{L}_C^{1,2}$ we define

$$D_t^+ u = \lim_{\epsilon \downarrow 0} D_t u_{t+\epsilon} \qquad D_t^- u = \lim_{\epsilon \downarrow 0} D_t u_{t-\epsilon}.$$

Then we have (cf. Theorem 7.3 of [6]):

**Proposition 2.2.** *If $u$ belongs to $\mathbb{L}_C^{1,2}$ then $u$ is Stratonovich integrable and*

$$\int_0^1 u_t \circ dW_t = \delta(u) + \frac{1}{2} \int_0^1 (D_t^+ u + D_t^- u) dt. \tag{2.5}$$

From the point of view of the stochastic calculus, the Stratonovich integral behaves as an ordinary pathwise integral. We refer to [6] for a detailed discussion of this fact.

The next proposition provides a different kind of sufficient conditions for the existence of the Stratonovich integral.

**Proposition 2.3.** *Let $u$ be a process in $\mathbb{L}_{loc}^{1,2}$. Suppose that the integral operator from $H$ into $H$ associated with the kernel $Du(\omega)$ is nuclear for all $\omega$ a.s. Then $u$ is Stratonovich integrable and we have*

$$\int_0^1 u_t \circ dW_t = \delta(u) + \text{Tr}\, Du. \tag{2.6}$$

*Proof:* From Proposition 6.1 of [9] we know that for any complete orthonormal system $\{e_i, i \geq 1\}$ in $H$ the series

$$\sum_{i=1}^{\infty} \left( \int_0^1 u_t e_i(t) dt \right) \int_0^t e_i(s) dW_s$$

converges in probability to $\delta(u) + \text{Tr}\, Du$. Then the proposition follows from the results of [10].                                                    Q.E.D.

We will finish this preliminary section with some comments about the noncausal Girsanov theorem. Let $\{u_t, 0 \leq t \leq 1\}$ be a measurable process such that $\int_0^1 u_t^2 dt < \infty$ a.s. Define the transformation $T : \Omega \longrightarrow \Omega$ by

$$T(\omega)_t = \omega_t + \int_0^t u_s(\omega) ds.$$

We know that if $u$ is adapted and if $E(J) = 1$, where $J = \exp\left(-\int_0^1 u_s dW_s - \frac{1}{2}\int_0^1 u_s^2 ds\right)$, then $\{T(\omega)_t\}$ is a Wiener process under a new probability $Q$ given by $dQ/dP = J$. For a non adapted process $u$ we need more restrictive conditions on the process in order to deduce a similar result. Let us first introduce the following definition (see Kusuoka [3]):

**Definition 2.4.** *Let $u$ be a measurable process such that $\int_0^1 u_t^2 dt < \infty$ a.s. We will say that $u$ is $\mathcal{H}$-$C^1$ (namely, the mapping $\omega \to u_\cdot(\omega)$ from $\Omega$ into $H$ is $\mathcal{H}$-$C^1$) if there exists a random kernel $Du(\omega) \in L^2([0,1]^2)$ such that:*
(i) $\|u(\omega + \int_0^\cdot h_s ds) - u(\omega) - Du(\omega)(h)\|_H = o(\|h\|_H)$ *for all $\omega \in \Omega$ as $\|h\|_H$ tends to zero.*

*(ii) The mapping $h \longmapsto Du(\omega + \int_0^\cdot h_s ds)$ is continuous from $H$ into $L^2([0,1]^2)$ for all $\omega$.*

One can prove (see the proof of Theorem 5.2 in Kusuoka [3]) that if $u$ is $\mathcal{H}-C^1$ then $u$ belongs to $\mathbb{L}_{loc}^{1,2}$, and the kernel $Du$ verifying the above conditions (i) and (ii) is precisely the operator $D$ applied to $u$. Furthermore if $u : \Omega \to H$ is continuously differentiable then $u$ is $\mathcal{H}-C^1$, and for every $\omega \in \Omega$, $Du(\omega) \in H \otimes H$ is the derivative of $u$. Using these defintions, we can state the following result proved by Kusuoka (see Theorem 6.4 of [3]).

**Theorem 2.5.** *Let $\{u_t, 0 \leq t \leq 1\}$ be a stochastic process which defines a $\mathcal{H}-C^1$ map from $\Omega$ into $L^2(0,1)$. Suppose that:*
*(i) The transformation $T : \Omega \to \Omega$ given by $T(\omega)_t = \omega_t + \int_0^t u_s(\omega) ds$ is bijective.*
*(ii) $I + Du(\omega) : H \to H$ is invertible, for all $\omega$ a.s.*
*Then the process $\{W_t + \int_0^t u_s ds\}$ is a Wiener process under the probability $Q$ on $C_0([0,1])$ given by*

$$\frac{dQ}{dP} = |d_c(-Du)| \exp\left(-\delta(u) - \frac{1}{2} \int_0^1 u_t^2 dt\right), \qquad (2.7)$$

*where $d_c(-Du)$ denotes the Carleman–Fredholm determinant of the square integrable kernel $Du \in L^2([0,1]^2)$.*

We refer to [15] and [17] for the definition and main properties of the Carleman–Fredholm determinant of a Hilbert–Schmidt operator. If the integral operator associated with the kernel $Du(\omega)$ is nuclear for each $\omega$, then the Carleman–Ferdholm determinant can be computed as follows (see [15]):

$$d_c(-Du) = \det(I + Du) \exp(-\text{Tr } Du). \qquad (2.8)$$

Moreover, from Proposition 2.3 we obtain in this case

$$\frac{dQ}{dP} = |\det(I + Du)| \exp\left(-\int_0^1 u_t \circ dW_t - \frac{1}{2} \int_0^1 u_t^2 dt\right). \qquad (2.9)$$

## 3. Second order stochastic differential equations with Dirichlet boundary conditions

In this section we are going to study the second order stochastic differential equation

$$\ddot{X}_t + f(X_t, t) = \dot{W}_t, \qquad 0 \leq t \leq 1, \qquad (3.1)$$

with the boundary conditions $X_0 = X_1 = 0$. Here $\{W_t\}$ is a one–dimensional Brownian motion starting at zero, and the function $f : \mathbb{R} \times [0,1] \to \mathbb{R}$, is supposed to be measurable and locally bounded. Equation (3.1) must be regarded as a formal differential version of the integral equation

$$\dot{X}_t + \int_0^t f(X_s, s) ds = \dot{X}_0 + W_t, \qquad 0 \leq t \leq 1. \qquad (3.2)$$

Before studying the equation (3.1) we will find the solution in the particular case $f = 0$. That means, let $\{Y_t\}$ be the solution of

$$\ddot{Y}_t = \dot{W}_t, \qquad 0 \le t \le 1,$$
$$Y_0 = Y_1 = 0. \tag{3.3}$$

The solution to the equation (3.3) is

$$Y_t = -t \int_0^1 W_s ds + \int_0^t W_s ds, \tag{3.4}$$

and we also have

$$\dot{Y}_t = W_t - \int_0^1 W_s ds. \tag{3.5}$$

Define the transformation $T : C_0([0,1]) \longrightarrow C_0([0,1])$ by

$$T(\omega)_t = \omega_t + \int_0^t f(Y_s(\omega), s) ds, \tag{3.6}$$

where $\{Y_t\}$ is given by (3.4). Notice that the mapping $\omega \longmapsto Y(\omega)$ from $C_0([0,1])$ into the space $C_{0,0}^1(0,1)$ of continuously differentiable functions $y$ on $(0,1)$ such that $\lim_{t\downarrow 0} y(t) = 0$ and $\lim_{t\uparrow 1} y(t) = 0$ is bijective.

We remark the following two facts:

(I) If $T(\eta) = W$, then the function $X_t = Y_t(\eta)$ is a solution of the equation (3.1). In fact, we have

$$\dot{X}_t = \dot{Y}_t(\eta) = -\int_0^1 \eta_s \, ds + \eta_t = \dot{X}_0 + W_t - \int_0^t f(X_s, s) \, ds.$$

(II) Conversely, if we are given a solution $\{X_t\}$ of equation (3.1), then $T(Y^{-1}(X)) = W$. Indeed, if we set $Y^{-1}(X) = \eta$, then

$$T(\eta)_t = \eta_t + \int_0^t f(Y_s(\eta), s) \, ds = \eta_t + W_t + \dot{X}_0 - \dot{X}_t = W_t.$$

Consequently, if $T$ is a *bijection* then equation (3.1) has the unique solution $X = Y(T^{-1}(W))$ for any continuous function $W \in C_0([0,1])$. In the sequel we will assume the following hypothesis:

(P.1): *The function $f$ is nonincreasing in the variable $x$, locally Lipschitz and with linear growth, uniformly with respect to the variable $t$. That means, for every $N > 0$ there exist constants $C > 0$ and $C_N > 0$ such that*

$$|f(x,t) - f(y,t)| \le C_N |x - y|, \qquad \forall \ |x|, |y| \le N, t \in [0,1],$$

$$|f(x,t)| \le C(1 + |x|), \qquad \forall x, t.$$

We will see in the next proposition that this hypothesis is sufficient for the existence and uniqueness of a solution of the equation (3.1).

**Proposition 3.1.** *Suppose that $f$ verifies (P.1). Then $T$ is a bijection and, consequently, there exists a unique solution to the equation (3.1) for any continuous function $W$ in $C_0([0,1])$.*

*Proof:* We need to show that for any $\eta \in C_0([0,1])$ there exists a unique function $W \in C_0([0,1])$ such that

$$\dot{\eta}_t = \dot{W}_t + f\left(-t\int_0^1 W_s ds + \int_0^t W_s ds, s\right).$$

Set $V = \eta - W$. Then $V$ satisfies

$$\dot{V}_t = f\left(t\int_0^1 V_s ds - \int_0^t V_s ds + \xi_t, t\right),$$

$$V_0 = 0,$$

where

$$\xi_t = -t\int_0^1 \eta_s ds + \int_0^t \eta_s ds.$$

For any $y \in \mathbb{R}$ we consider the differential equation

$$\dot{V}_t(y) = f\left(ty - \int_0^t V_s(y) ds + \xi_t, t\right)$$

$$V_0(y) = 0.$$

By a comparison theorem for ordinary differential equations and using the monotonicity properties of $f$ we deduce that the mapping $y \longmapsto V_t(y)$ is continuous and nonincreasing for each $t \in [0,1]$. Therefore, $\int_0^1 V_t(y) dt$ is a nonincreasing and continuous function of $y$, and this implies the existence of a unique real number $y$ such that $\int_0^1 V_t(y) dt = y$. This completes the proof of the proposition.                                                 Q.E.D.

We want to study the Markov properties of the process $\{X_t\}$ solution of the equation (3.1). Let us first recall the following types of Markov property:

(1) We say that a $d$–dimensional stochastic process $\{Z_t, 0 \le t \le 1\}$ is a *Markov process* if for any $t \in [0,1]$ the past and the future of $\{Z_s\}$ are conditionally independent, given the present state $Z_t$.

(2) We say that $\{Z_t, 0 \le t \le 1\}$ is a *Markov field* if for any $0 \le s < t \le 1$, the values of the process inside and outside the interval $[s,t]$ are conditionally independent, given $Z_s$ and $Z_t$.

Following the results of Russek [14] we might conjecture that as a solution of a second order stochastic differential equation the process $\{X_t\}$ is a 2–Markov process, that means, the two dimensional process $\{(X_t, \dot{X}_t)\}$ is a Markov process. We first show that this is true for the process $\{Y_t\}$ i.e., when $f \equiv 0$.

**Proposition 3.2.** *The process $\{(Y_t, \dot{Y}_t), 0 \le t \le 1\}$ defined by the equations (3.4) and (3.5) is a Markov process.*

*Proof:* Let $\psi(x,y)$ be a real valued bounded and measurable function. Fix $s < t$ and set $\zeta = \int_0^1 W_t\, dt$. We have to compute the conditional expectation

$$E(\psi(Y_t, \dot{Y}_t)|(Y_r, \dot{Y}_r),\ 0 \le r \le s)$$

$$= E(\psi(-t\zeta + \int_0^t W_u du,\ -\zeta + W_t)\,|\,\zeta,\ W_r,\ 0 \le r \le s)$$

$$= E(\psi(-t\zeta + \int_0^s W_u du + \int_s^t (W_u - W_s)du + (t - s)W_s,$$
$$-\zeta + W_t - W_s + W_s)|\zeta,\ W_r,\ 0 \le r \le s)$$

$$= \int_{\mathbb{R}^2} \psi\Big(-t\zeta + \int_0^s W_u\, du + x + (t - s)W_s,\ -\zeta + y + W_s\Big)$$

$$\cdot N\Big(\Big(\frac{(t-s)^2(3-2s-t)}{2(1-s)^3}\int_s^1 (W_u - W_s)\, du,$$

$$\frac{3(t-s)(2-s-t)}{2(1-s)^3}\int_s^1 (W_u - W_s)\, du\Big),\ \wedge\Big)(dx, dy)$$

where $\wedge$ denotes the conditional covariance matrix of the Gaussian vector $(\int_s^t (W_u - W_s)\, du,\ W_t - W_s)$, given $\int_s^1 (W_t - W_s)dt$. Consequently, the above conditional expectation will be a function of the random variables

$$-t\zeta + \int_0^s W_u\, du + (t - s)\, W_s = (t - s)\dot{Y}_s + Y_s,$$

$$-\zeta + W_s = \dot{Y}_s,$$

$$\int_s^1 (W_u - W_s)\, du = -(1 - s)\dot{Y}_s - Y_s,$$

and this implies the Markov property. Q.E.D.

We will see that, except in the linear case, the Markov property does not hold for the process $\{X_t\}$. One might think that the Markov field property is better adapted to our equation because we impose fixed values at the boundary points $t = 0$ and $t = 1$. However this is not the case and, as we shall see, the nonlinearity of the function $f$ prevents for any type of Markov property. The main result of this section is the following.

**Theorem 3.3.** *Let $\{X_t, t \in [0,1]\}$ be the solution of equation (3.1) and suppose that the function $f$ has linear growth, $f$ is of class $C^2$ with respect to $x$, its partial derivatives $f_x'$, $f_{xx}''$ are continuous in $(x,t)$, and $f_x' \le 0$. Then if $f$ is an affine function of $x$, $\{(X_t, \dot{X}_t)\}$ is a Markov process, and conversely, if this process is a Markov field, then $f_{xx}'' \equiv 0$.*

In order to prove this theorem we are going to introduce a new probability measure $Q$ on $C_0([0,1])$ such that $P = Q \circ T^{-1}$, where $T$ is the mapping defined by (3.6). Then

$\{T(\omega)_t\}$ will be a Wiener process uner $Q$, and, consequently, the law of the process $\{X_t\}$ under the probability $P$ coincides with the law of $\{Y_t\}$ under $Q$. In fact, we have

$$P\{\omega : X(\omega) \in B\} = Q\{\omega : X(T(\omega)) \in B\} = Q\{\omega : Y(\omega) \in B\},$$

for any Borel subset $B$ of $C^1_{0,0}([0,1])$. In this way we will translate the problem of the Markov property of $\{X_t\}$ into the problem of the Markov property of the process $\{Y_t\}$ under a new probability $Q$. This problem can be handled, provided $Q$ is absolutely continuous with respect to the Wiener measure $P$ and we can compute an explicit expression for its Radon–Nikodym derivative. To do this we will make use of Theorem 2.5. So, before proving Theorem 3.3 we will present some preliminary lemmas. Let us first introduce some notations. Set $\alpha_t = f'_x(Y_t, t)$, and denote by $M_t$ the matrix $\begin{bmatrix} 0 & -\alpha_t \\ 1 & 0 \end{bmatrix}$. Let $\Phi_t$ be the solution of the linear differential equation

$$d\Phi_t = M_t \Phi_t \, dt$$
$$\Phi_0 = I \tag{3.6}$$

We will also denote by $\Phi(t,s)$ the matrix $\Phi_t \Phi_s^{-1}$.

**Lemma 3.4.** *Let $f$ be a function satisfying the conditions of Theorem 3.3. Then the process $u_t = f(Y_t, t)$ verifies the conditions of the generalized Girsanov Theorem 2.5 and we have*

$$\frac{dQ}{dP} = Z_1 \exp\left(-\int_0^1 f(Y_t, t) \circ dW_t - \frac{1}{2}\int_0^1 f(Y_t, t)^2 \, dt\right), \tag{3.7}$$

*where $Z_1$ is the solution at time $t = 1$ of the second order differential equation*

$$\begin{cases} \ddot{Z}_t + \alpha_t Z_t = 0 \\ Z_0 = 0, \dot{Z}_0 = 1. \end{cases} \tag{3.8}$$

*Proof of Lemma 3.4.* First we have to check that the process

$$u_t = f(Y_t, t) = f\left(-t\int_0^1 W_s ds + \int_0^t W_s ds, t\right)$$

satisfies the conditions of Theorem 2.5. We already know that the mapping $T(\omega)_t = \omega_t + \int_0^t u_s(\omega)ds$ is bijective, due to Proposition 3.1. Furthermore, $u$ is $\mathcal{H}$–$C^1$ because the mapping $\omega \longmapsto u_\cdot(\omega)$ is continuously differentiable from $\Omega$ into $H$. So, it remains to show that $I + Du$ is invertible a.s. From the properties of the operator $D$ we deduce that

$$D_s Y_t = -t(1-s) + (t-s)^+ = st - s \wedge t, \tag{3.9}$$

and therefore,

$$D_s u_t = \alpha_t(st - s \wedge t). \tag{3.10}$$

From the Fredholm alternative, in order to show that $I + Du$ is invertible, it suffices to check that $-1$ is not an eigenvalue of $Du(\omega)$, for each $\omega \in \Omega$ a.s. Let $h \in L^2(0,1)$ such that $(I + Du)h = 0$. Then

$$h_t + \alpha_t \int_0^1 h_s(st - s \wedge t)\,ds = 0,$$

which can be written as

$$h_t + t\alpha_t \int_0^1 s\,h_s\,ds - \alpha_t \int_0^t s\,h_s\,ds - t\alpha_t \int_t^1 h_s\,ds = 0.$$

Setting $g_t = \int_0^t h_s\,ds$ and $U_t = \int_0^t g_s\,ds$ we obtain

$$\ddot{U}_t + \alpha_t U_t - t\alpha_t U_1 = 0. \tag{3.11}$$

Then the solution of the second order differential equation (3.11) is given by

$$U_t = U_1 \int_0^t \Phi_{21}(t,s)s\alpha_s\,ds, \tag{3.12}$$

which implies $U_1 = 0$ because $\Phi_{21} \geq 0$ and $\alpha_s \leq 0$.

To finish the proof of the lemma it remains to establish the formula (3.7). To do this we have to compute the Carleman–Fredholm determinant of the kernel (3.10). First observe that this kernel is nuclear and continuous. In fact, the cointinuity is clear. On the other hand, the kernel $st\alpha_t$ is nuclear. Now $\lambda_n = 4/[(2n-1)^2\pi^2]$, $n \geq 1$ are the eigenvalues of the integral operator associated with the kernel $\min(s,t)$ and $e_n(t) = \sqrt{2}\sin(2n-1)\frac{\pi t}{2}$ the corresponding eigenfunctions. Then

$$\int_0^1 \alpha_t(s \wedge t)e_n(s)ds = \alpha_t \lambda_n e_n(t),$$

and

$$\sum_{n=1}^{\infty} \|\alpha_t \lambda_n e_n(t)\|_2 \leq \|\alpha\|_\infty \sum_{n=1}^{\infty} \lambda_n < \infty,$$

which implies the nuclearity of the kernel $\alpha_t(s \wedge t)$. Therefore we can make use of the formula (2.9), and it suffices to show that $\det(I + Du)$ is equal to $Z_1$. Note that the absolute value of $Z_1$ can be omitted because $Z_t \geq 0$. Set $k(s,t) = st - s \wedge t$. From a well–known formula for the determinant of a nuclear operator (see e.g. [15]) we obtain

$$\det(I + Du) = \sum_{n=0}^{\infty} \frac{\gamma_n}{n!},$$

where

$$\gamma_n = \int_{[0,1]^n} \det(D_{t_i} u_{t_j})dt_1 \cdots dt_n.$$

We have to compute the coefficients $\gamma_n$.

$$\gamma_n = \int_{[0,1]^n} \det(k(t_i,t_j))\alpha(t_1)\cdots\alpha(t_n)dt_1\cdots dt_n$$

$$= n! \int_{\{t_1<t_2<\cdots<t_n\}} \det(k(t_i,t_j))\alpha(t_1)\cdots\alpha(t_n)dt_1\cdots dt_n.$$

The determinant of the matrix $(k(t_i,t_j))$ is equal to

$$\left(\prod_{i=1}^n t_i\right)^2 \det \begin{bmatrix} 1-\frac{1}{t_1} & 1-\frac{1}{t_2} & 1-\frac{1}{t_3} & \cdots & 1-\frac{1}{t_n} \\ 1-\frac{1}{t_2} & 1-\frac{1}{t_2} & 1-\frac{1}{t_3} & \cdots & 1-\frac{1}{t_n} \\ 1-\frac{1}{t_3} & 1-\frac{1}{t_3} & 1-\frac{1}{t_3} & \cdots & 1-\frac{1}{t_n} \\ \cdots & \cdots & \cdots & & \cdots \\ 1-\frac{1}{t_n} & 1-\frac{1}{t_n} & 1-\frac{1}{t_n} & \cdots & 1-\frac{1}{t_n} \end{bmatrix}.$$

Substracting the $k$–th row from the $(k-1)$–th one, $k=2,\dots,n$, we get

$$\det(k(t_i,t_j)) = \left(\prod_{i=1}^n t_i\right)^2$$

$$\times \left(\frac{1}{t_2}-\frac{1}{t_1}\right)\left(\frac{1}{t_3}-\frac{1}{t_2}\right)\left(\frac{1}{t_4}-\frac{1}{t_3}\right)\cdots\left(\frac{1}{t_n}-\frac{1}{t_{n-1}}\right)\left(1-\frac{1}{t_n}\right)$$

$$= (-1)^n t_1(t_2-t_1)(t_3-t_2)(t_4-t_3)\cdots(t_n-t_{n-1})(1-t_n).$$

Consequently we obtain

$$\det(I+Du) = \sum_{n=0}^{\infty}(-1)^n$$

$$\times \int_{\{t_1<t_2<\cdots<t_n\}} t_1(t_2-t_1)(t_3-t_2)\cdots(t_n-t_{n-1})(1-t_n)\alpha(t_1)\cdots\alpha(t_n)dt_1\cdots dt_n$$

$$(3.13)$$

On the other hand the solution of the second order differential equation (3.8) can be expressed as

$$Z_t = t - \int_0^t \int_0^s \alpha_u Z_u du = t - \int_0^t (t-u)\alpha_u Z_u du.$$

Iterating this equality we deduce that $\det(I+Du)$ coincides with $Z_1$. The proof of the lemma is now complete.                                                          Q.E.D.

**Lemma 3.5.** *Let $\mathcal{G}_t$ be the $\sigma$–algebra generated by $Y_t$, $\dot{Y}_t$ and $\int_0^1 W_s\,ds$, where we recall that $Y_t$ and $\dot{Y}_t$ are defined by (3.4) and (3.5). Let $F$ be a random variable in the space $\mathbb{D}_{loc}^{1,2}$ such that $F\,1_G$ is $\mathcal{G}_t$–measurable for some set $G \in \mathcal{G}_t$. Then there exist random variables $A_t$, $B_t$, and $C_t$ such that*

$$1_G D_\theta F = 1_G\left[A_t\theta + C_t\right]1_{[0,t]}(\theta) + 1_G B_t(\theta-1)1_{[t,1]}(\theta),$$

*for $dP \times d\theta$ almost all $(\omega,\theta) \in \Omega \times [0,1]$.*

*Proof:* Let us first compute

$$D_\theta Y_t = t(1-\theta) + (t-\theta)\mathbf{1}_{[0,t]}(\theta) = \theta(t-1)\mathbf{1}_{[0,t]}(\theta) + (\theta-1)t\,\mathbf{1}_{[t,1]}(\theta)\,, \qquad (3.14)$$

$$D_\theta \dot{Y}_t = -(1-\theta) + \mathbf{1}_{[0,t]}(\theta) = \theta\,\mathbf{1}_{[0,t]}(\theta) + (\theta-1)\mathbf{1}_{[t,1]}(\theta)\,, \qquad (3.15)$$

and

$$D_\theta\Big(\int_0^1 W_t\,dt\Big) = 1 - \theta\,. \qquad (3.16)$$

Thus, it suffices to apply Lemma 2.1 to the three–dimensional subspace of $H$ generated by $\theta\mathbf{1}_{[0,t]}(\theta)$, $\mathbf{1}_{[0,t]}(\theta)$ and $(\theta-1)\mathbf{1}_{[t,1]}(\theta)$. \hfill Q.E.D.

*Proof of Theorem 3.3.* Let $Q$ be the probability measure on $C_0([0,1])$ given by Lemma 3.4. The law of the process $\{X_t\}$ under $P$ is the same as the law of $\{Y_t\}$ under $Q$. Therefore, we can replace the process $\{(X_t, \dot{X}_t)\}$ by $\{(Y_t, \dot{Y}_t)\}$ and the probability $P$ by $Q$ in the statement of the theorem. By Proposition 3.2 $\{(Y_t, \dot{Y}_t)\}$ is a Markov process under $P$ and now we have to study the Markov property with respect to an equivalent probability measure $Q$. The Radon–Nikodym derivative $J = \frac{dQ}{dP}$ will be given by the formula (3.7). Notice also that $Z_t > 0$ for all $t$ in $(0,1]$, and $\dot{Z}_t > 0$ for all $t$ in $[0,1]$. For any fixed $t \in (0,1)$ we can factorize the Radon–Nikodym derivative $J = \frac{dQ}{dP}$ as follows

$$J = Z_1 L_t L^t,$$

where

$$L_t = \exp\left(-\int_0^t f(Y_s, s) \circ dW_s - \frac{1}{2}\int_0^t f(Y_s, s)^2\,ds\right),$$

$$L^t = \exp\left(-\int_t^1 f(Y_s, s) \circ dW_s - \frac{1}{2}\int_t^1 f(Y_s, s)^2\,ds\right).$$

We define the $\sigma$–algebras
$\mathcal{F}_t = \sigma\{(Y_s, \dot{Y}_s),\ 0 \le s \le t\}$
$\mathcal{F}^t = \sigma\{(Y_s, \dot{Y}_s),\ t \le s \le 1\}$, and $\mathcal{F}_0^t = \mathcal{F}^t \vee \sigma\{Y_0, \dot{Y}_0\} = \mathcal{F}^t \vee \sigma\{\int_0^1 W_t\,dt\}$.
For any random variable $\xi$ integrable with respect to $Q$ we set

$$\Lambda_\xi = E_Q(\xi|\mathcal{F}_t) = \frac{E_P(\xi J|\mathcal{F}_t)}{E_P(J|\mathcal{F}_t)} = \frac{E_P(\xi Z_1 L^t|\mathcal{F}_t)}{E_P(Z_1 L^t|\mathcal{F}_t)}, \qquad (3.17)$$

because $L_t$ is $\mathcal{F}_t$–measurable.

(i) Suppose first that $f$ is an affine function. In that case $Z_1$ is deterministic and we get

$$\Lambda_\xi = \frac{E_P(\xi L^t|\mathcal{F}_t)}{E_P(L^t|\mathcal{F}_t)}\,.$$

Then if $\xi$ is $\mathcal{F}^t$–measurable, using the fact that $L^t$ is also $\mathcal{F}^t$–measurable and applying the Markov property of $\{(Y_t, \dot{Y}_t)\}$ under $P$ we deduce that $\Lambda_\xi$ is $\sigma\{Y_t, \dot{Y}_t\}$–measurable and this implies that $\{(Y_t, \dot{Y}_t);\ t \in [0,1]\}$ is a Markov process under $Q$.

(ii) To prove the converse assertion we suppose that $\{(Y_t, \dot{Y}_t); t \in [0,1]\}$ is a Markov field under $Q$ and we have to show that $f''_{xx} = 0$. This assumption implies in particular that for any $t \in (0,1)$ and any $\mathcal{F}^t_0$-measurable random variable $\xi$, integrable with respect to $Q$, the conditional expectation $\Lambda_\xi = E_Q(\xi|\mathcal{F}_t)$ is $\mathcal{G}_t = \sigma\{Y_t, \dot{Y}_t, \int_0^1 W_t\, dt\}$-measurable. In order to obtain a suitable factorization of the random variable $Z_1$ we set

$$\begin{bmatrix} \dot{Z}_1 \\ Z_1 \end{bmatrix} = \Phi(1,t)\begin{bmatrix} \dot{Z}_t \\ Z_t \end{bmatrix}, \tag{3.18}$$

that means,

$$Z_1 = \Phi_{21}(1,t)\dot{Z}_t + \Phi_{22}(1,t)Z_t, \tag{3.19}$$

where the process $\Phi(t)$ has been introduced in (3.6). Now define

$$\varphi_t = \frac{\Phi_{21}(1,t)}{\Phi_{22}(1,t)} \quad \text{and} \quad \psi_t = \frac{Z_t}{\dot{Z}_t} = \frac{\Phi_{21}(t)}{\Phi_{11}(t)}. \tag{3.20}$$

From the equation $\widehat{\Phi^{-1}}(t) = -\Phi^{-1}(t)M_t$, we deduce that $\Phi_{21}(1,t)$ sastisfies the same second order differential equation (3.8) as $Z_t$. From this fact we deduce that $\varphi_t$ and $\psi_t$ are continuously differentiable processes on $[0,1]$, $\varphi_1 = 0$, $\psi_0 = 0$, $\varphi_t > 0$ for $t \in [0,1)$ and $\psi_t > 0$ for $t \in (0,1]$. Then we can write

$$Z_1 = \Phi_{22}(1,t)\dot{Z}_t(\varphi_t + \psi_t), \tag{3.21}$$

and we obtain the factorization

$$J = L_t\dot{Z}_tC^t(\varphi_t + \psi_t), \tag{3.22}$$

where $C^t = L^t\Phi_{22}(1,t)$.

In the sequel we will denote by $\overline{F}$ the conditional expectation of the random variable $F$ under $P$ with respect to the $\sigma$-algebra $\mathcal{G}_t$. The random variables $\dot{Z}_t$ and $\psi_t$ are $\mathcal{F}_t$-measurable and, on the other hand, $\Phi_{22}(1,t)$, and $\varphi_t$ are $\mathcal{F}^t$-measurable. Thus, from (3.17) and (3.22) and applying the Markov field property of $\{(Y_t, \dot{Y}_t)\}$ under $P$ we deduce that

$$\Lambda_\xi = \frac{\overline{\xi C^t}\psi_t + \overline{\xi C^t\varphi_t}}{\overline{C^t}\psi_t + \overline{C^t\varphi_t}}$$

and by our hypotheses this expression is $\mathcal{G}_t$-measurable. Therefore we obtain the following equation

$$\left(\Lambda_\xi\, \overline{C^t} - \overline{\xi C^t}\right)\psi_t + \left(\Lambda_\xi\, \overline{C^t\varphi_t} - \overline{\xi C^t\varphi_t}\right) = 0, \tag{3.23}$$

which is valid for any $\mathcal{F}^t_0$-measurable random variable $\xi$ integrable with respect to $Q$. Now we choose two particular variables $\xi$:

$$\xi_1 = (C^t)^{-1} \quad \text{and} \quad \xi_2 = (C^t\varphi_t)^{-1}.$$

First we remark that for $t \in (0,1)$, $i = 1,2$, $\xi_i$ is $Q$-integrable and nonnegative. In fact, this follows easily from the estimates $0 \le \Phi_{21}(1,t)^{-1} \le (1-t)^{-1}$ and $0 \le \Phi_{22}(1,t)^{-1} \le 1$.

Then we define the set

$$G_t = \left\{ \Lambda_{\xi_1} \overline{C^t \varphi_t} = \overline{\xi_1 C^t \varphi_t} \right\} \cap \left\{ \Lambda_{\xi_2} \overline{C^t \varphi_t} = \overline{\xi_2 C^t \varphi_t} \right\}.$$

Notice that $G_t \in \mathcal{G}_t$. On the set $G_t$ we have

$$\Lambda_{\xi_1} = \frac{\overline{\xi_1 C^t}}{\overline{C^t}} = \frac{\overline{\xi_1 C^t \varphi_t}}{\overline{C^t \varphi_t}},$$

$$\Lambda_{\xi_2} = \frac{\overline{\xi_2 C^t}}{\overline{C^t}} = \frac{\overline{\xi_2 C^t \varphi_t}}{\overline{C^t \varphi_t}}.$$

Consequently

$$\overline{\varphi_t} \overline{C^t} = \overline{C^t \varphi_t}$$

$$\overline{\varphi_t^{-1} C^t \varphi_t} = \overline{C^t},$$

which implies that $\left( \overline{\varphi_t} \right)^{-1} = \overline{\varphi_t^{-1}}$. By the strict Jensen inequality applied to the measure space $(G_t, \mathcal{F}|_{G_t}, P)$ we get that the random variable $1_{G_t} \varphi_t$ is $\mathcal{G}_t$-measurable. From the equation (3.23) we deduce that the random variable $1_{G_t^c} \psi_t$ is $\mathcal{G}_t$-measurable.

It is not hard to show that the random variables $\varphi_t$ and $\psi_t$ belong to the space $\mathbb{D}_{\text{loc}}^{1,2}$, for any $t \in (0,1)$. Thus by Lemma 3.5 applied to the random variables $\varphi_t$ and $\psi_t$ and to the sets $G_t$ and $G_t^c$, there exist random variables $\Gamma_1(t)$ and $\Gamma_2(t)$ such that

$$D_\theta \varphi_t = (\theta - 1) \Gamma_1(t), \tag{3.24}$$

for all $\theta \in [t,1]$, $\omega \in G_t$, a.e., and

$$D_\theta \psi_t = \theta \Gamma_2(t), \tag{3.25}$$

for all $\theta \in [0,t]$, $\omega \in G_t c$, a.e.

On the other hand, from the linear differential equations satisfied by $\Phi(t)$ and $\Phi(1,t)$ we can derive Ricatti type differential equations for $\varphi_t$ and $\psi_t$. In fact, differentiating with respect to $t$ the equations

$$\Phi_{21}(1,t) = \varphi_t \Phi_{22}(1,t)$$

$$\Phi_{21}(t) = \psi_t \Phi_{11}(t),$$

we obtain

$$\dot{\varphi}_t + \alpha_t \varphi_t^2 + 1 = 0; \quad \varphi_1 = 0 \tag{3.26}$$

$$\dot{\psi}_t - \alpha_t \psi_t^2 - 1 = 0; \quad \psi_0 = 0 \tag{3.27}$$

Applying the operator $D$, which commutes with the derivative with respect to the time variable, to the equations (3.26) and (3.27) yields

$$\frac{d}{dt} D_\theta \varphi_t + 2\varphi_t \alpha_t D_\theta \varphi_t + \varphi_t^2 D_\theta \alpha_t = 0, \quad D_\theta \varphi_1 = 0,$$

$$\frac{d}{dt} D_\theta \psi_t - 2\psi_t \alpha_t D_\theta \psi_t - \psi_t^2 D_\theta \alpha_t = 0, \quad D_\theta \psi_0 = 0. \tag{3.28}$$

Set $\gamma_{ts} = \exp\left(\int_t^s 2\varphi_r\alpha_r dr\right)$, for any $s, t \in [0, 1]$. Notice that

$$D_\theta\alpha_t = \theta(t-1)f_{xx}''(Y_t, t)\mathbf{1}_{[0,t]}(\theta) + (\theta-1)tf_{xx}''(Y_t, t)\mathbf{1}_{[t,1]}(\theta) . \qquad (3.29)$$

Then if we solve the linear equation satisfied for $D_\theta\varphi_t$ as a function of $t$, starting from the point 1 we get

$$
\begin{aligned}
D_\theta\varphi_t &= \int_t^1 \gamma_{ts}\varphi_s^2 D_\theta\alpha_s ds \\
&= (\theta-1)\int_t^\theta \gamma_{ts}f_{xx}''(Y_s, s)s\,ds + \theta\int_\theta^1 \gamma_{ts}f_{xx}''(Y_s, s)(s-1)ds \qquad (3.30) \\
&= (\theta-1)\int_t^1 \gamma_{ts}f_{xx}''(Y_s, s)s\,ds + \theta\int_\theta^1 \gamma_{ts}f_{xx}''(Y_s, s)(s-\theta)ds
\end{aligned}
$$

On the set $G_t$ and for $\theta \in [t, 1]$ we know by (3.24) that $D_\theta\varphi_t$ must be a multiple of $(\theta-1)$. Taking into account the equation (3.30), this is only possible if $f_{xx}''(Y_s, s) = 0$ for all $s$ in the interval $[t, 1]$ and for all $\omega \in G_t$ a.e. A similar argument, using the process $\psi_t$, yields that $f_{xx}''(Y_s, s) = 0$ for all $s$ in $[0, t]$ and for all $\omega \in G_t^c$ a.e. The point $t$ being arbitrary in $(0, 1)$ we deduce that the process $f_{xx}''(Y_s, s)$ is identically zero. Therefore $f_{xx}''(x) = 0$ for all $x$, which completes the proof of the theorem.          Q.E.D.

## Remark

(1) A similar result for a second order differential equation of the form

$$\ddot{X}_t + f(X_t, \dot{X}_t) = \dot{W}_t$$

has been proved in [8]. The proof in this case is more involved and some additional smoothness conditions on the function $f$ are required.

(2) As in [8] one could show that under the conditions of Theorem 3.3, if $\{(X_t, \dot{X}_t)\}$ is a germ Markov field then $f_{xx}'' = 0$. We recall that $\{(X_t, \dot{X}_t), 0 \le t \le 1\}$ is a *germ Markov field* if for any $0 \le s < t \le 1$, the values of the process inside and outside the interval $[s, t]$ are conditionally independent, given the germ $\sigma$–field $\bigcap_{\epsilon>0} \sigma((X_u, \dot{X}_u), u \in (s-\epsilon, s+\epsilon) \cup (t-\epsilon, t+\epsilon))$.

## 4. First order stochastic differential equations with boundary conditions

In this section we describe some particular types of stochastic differential equations that have been recently studied using the techniques of noncausal stochastic calculus.

*4.1 Bilinear stochastic differential equations of Stratonovich type with boundary conditions.*

In [11] Ocone and Pardoux have developed a detailed study of an equation of the form

$$
\begin{aligned}
dX_t &= (FX_t + a)dt + \sum_{i=1}^k (G_iX_t + b_i) \circ dW_t^i, \qquad 0 \le t \le 1 \\
H_0X_0 &+ H_1X_1 = h_0.
\end{aligned}
\qquad (4.1)
$$

Here $F$, $G_i$, $H_0$ and $H_1$ are $d \times d$ matrices, and $a$, $b_i$ and $h_0$ are vectors in $\mathbb{R}^d$. We assume that the matrix $[H_0 : H_1]$ has rank $d$. Particular examples of this type of boundary conditions are:

(1) The first $l$ coordinates of $X_0$ are given, where $1 \le l \le d$, and the last $d - l$ coordinates of $X_1$ are also given. The above linear boundary conditions can be reduced to this situation whenever $\mathrm{Im}H_0 \cap \mathrm{Im}H_1 = \{0\}$.

(2) We impose the periodic boundary condition $X_0 = X_1$.

In order to find a solution for the equation (4.1) we introduce the fundamental solution of the corresponding linear system:

$$d\Phi_t = F\Phi_t dt + \sum_{i=1}^{k} G_i \Phi_t \circ dW_t^i, \qquad \Phi_0 = I,$$

and we set $V_t = at + \sum_{i=1}^{k} b_i W_t^i$. Then the candidate for a solution is

$$X_t = \Phi_t X_0 + \Phi_t \int_0^t \Phi_s^{-1} \circ dV_s. \tag{4.2}$$

Substituting (4.2) into the boundary condition we obtain

$$X_t = \Phi_t \left[ (H_0 + H_1 \Phi_1)^{-1} \left( h_0 - H_1 \Phi_1 \int_0^1 \Phi_s^{-1} \circ dV_s \right) \right] + \Phi_t \int_0^t \Phi_s^{-1} \circ dV_s. \tag{4.3}$$

One can show that (4.3) is the unique solution in some space of processes $\mathbb{L}_{\mathrm{loc}}^S$, provided that $\det(H_0 + H_1 \Phi_1) \neq 0$ a.s. We refer the reader to [11] for more details about this class of processes and for the proof of the existence and uniqueness of a solution.

Concerning the Markov property of the solution one can show the following results:

**Proposition 4.1.** *The solution $\{X_t\}$ of the equation (4.1) is a Markov process under one of the following situations:*
*(i) $H_1 = 0$ or $H_0 = 0$.*
*(ii) $G_i = 0$, $1 \le i \le k$ (Gaussian case) and $\mathrm{Im}H_0 \cap \mathrm{Im}H_1 = \{0\}$.*

If $\mathrm{Im}H_0 \cap \mathrm{Im}H_1 \neq \{0\}$ the process $X$ may not be Markovian even in the Gaussian case. On the other hand, in the non Gaussian case, the condition $\mathrm{Im}H_0 \cap \mathrm{Im}H_1 = \{0\}$ does not insure the Markov property. Concerning the Markov field property one has the following result.

**Proposition 4.2.** *The solution $\{X_t\}$ of the equation (4.1) is a Markov field in the following cases:*
*(i) $G_1 = \cdots = G_k = 0$ (Gaussian case)*
*(ii) $a = b_1 = \cdots = b_k = 0$, and $\Phi_t$ is a diagonal matrix for all $t \in [0, 1]$ (For instance, this is true in dimension one)*

*4.2. Non linear stochastic differential equations with boundary conditions.*

Consider the following equation

$$dX_t + f(X_t) = dW_t, \qquad 0 \le t \le 1$$
$$h(X_0, X_1) = h_0, \tag{4.4}$$

where $W$ is a $d$–dimensional Wiener process, $f : \mathbb{R}^d \to \mathbb{R}^d$, $h : \mathbb{R}^{2d} \to \mathbb{R}^d$, and $h_0$ is a vector in $\mathbb{R}^d$. The existence and uniqueness of a solution for the equation (4.4) has been discussed in [7]. The Markov property of the process $X$ has also been investigated using the techniques of Section 3. Concerning the Markov property one also has negative results in dimension one when the function $f$ is non linear. Let us describe briefly how one can obtain these results.

Notice first that if we assume a periodic boundary conditions, then the equation (4.4) with $f = 0$ has no solution. For this reason we introduce a decomposition of the form $f(x) = Ax + \overline{f}(x)$, where $A$ is a fixed deterministic $d \times d$–matrix. Then we denote by $Y$ the solution of the corresponding boundary value problem in the case $\overline{f} \equiv 0$. That means

$$dY_t + AY_t = dW_t,$$
$$h(Y_0, Y_1) = h_0 \tag{4.5}$$

The boundary value problem (4.5) is equivalent to the following system

$$Y_t = e^{-At} \left( Y_0 + \int_0^t e^{As} dW_s \right)$$
$$h \left( Y_0, e^{-A} \left( Y_0 + \int_0^1 e^{As} dW_s \right) \right) = h_0. \tag{4.6}$$

In order to solve the equation (4.6) we will impose the following assumption:

(H.1) *For all $z \in \mathbb{R}^d$ the equation $h(y, e^{-A}(y + z)) = h_0$ has a unique solution $y = g(z)$.*

We denote by $\Omega$ the space of continuous functions from $[0, 1]$ into $\mathbb{R}^d$ which vanish at zero. On the other hand we denote by $\Sigma$ the space of continuous functions $x$ from $[0, 1]$ into $\mathbb{R}^d$ which satisfy the desired boundary condition, that means $h(x_0, x_1) = h_0$. Then, under the hypothesis (H.1) there exists a bijection $\Psi : \Omega \longrightarrow \Sigma$ such that $Y = \Psi(W)$. We define the mapping $T : \Omega \longrightarrow \Omega$ by

$$T(\omega)_t = \omega_t + \int_0^t \overline{f}(\Psi_s(\omega)) ds.$$

As in Section 3 one can show that if $T$ is a bijection then the equation (4.4) has the unique solution $X = \Psi(T^{-1}(W))$. The following are sufficient conditions for the transformation $T$ to be bijective:

(1) *$T$ is one-to-one if:*

   (i) There exists $\lambda \in \mathbb{R}$ such that $f + \lambda I$ is strictly monotone and
   (ii) $e^{\lambda}|g(z) - g(z')| \le |e^{-A}(z - z' + g(z) - g(z'))|$, for all $z, z' \in \mathbb{R}^d$.

(2) $T$ is *onto* if:
  (iii) $\overline{f}$ is locally Lipschitz
  (iv) $\lim_{a \to \infty} \frac{1}{a} \sup_{|x| \le a} |\overline{f}(x)| = 0$    (sublinear growth)
  (v) $g$ has linear growth

Assuming (i), a sufficient condition for (ii) is the following:
  (ii') $h(x,y) = h(\bar{x}, \bar{y}) = h_0$   implies $e^\lambda |x - \bar{x}| \le |y - \bar{y}|$.

We remark that in the periodic case we only need $f$ to be strictly monotone for $T$ to be one–to–one, because we can take $\lambda = 0$ in (1).

We can now study the Markov property of the process $X$ solution of (4.4). By means the extended version of Girsanov theorem (see Theorem 2.5) one can show the following result:

**Theorem 4.3.** *Suppose that $\overline{f}$ and $g$ are continuously differentiable functions such that the mapping $T$ is bijective. Suppose in addition that*

$$\det \left[ I - e^A \Phi_1 g'(\xi_1) + g'(\xi_1) \right] \neq 0, \tag{4.7}$$

*where $\xi_t = \int_0^t e^{As} dW_s$, and $\Phi_t = I - \int_0^t f'(Y_s) \Phi_s ds$. Then the conditions of Theorem 2.5 are satisfied by the transformation $T$, and we have*

$$\frac{dQ}{dP} = \left| \det \left[ I - e^A \Phi_1 g'(\xi_1) + g'(\xi_1) \right] \right|$$

$$\times \exp \left\{ \int_0^t \mathrm{Tr} \overline{f}'(Y_t) dt - \int_0^t \overline{f}(Y_t) \circ dW_t - \frac{1}{2} \int_0^t |\overline{f}(Y_t)|^2 dt \right\}.$$

Theorem 4.3 can be used to investigate the Markov properties of the process $X$. In fact, setting $\overline{W}_t = W_t + \int_0^t \overline{f}(Y_s) ds$, we obtain

$$dY_t + AY_t = dW_t = d\overline{W}_t - \int_0^t \overline{f}(Y_s) ds.$$

Therefore, the process $\{Y_t\}$ is a solution of the equation (4.4) for the process $\overline{W}$, and this implies that the law of $Y$ under the probability $Q$ coincides with the law of $X$ under $P$. The results one can obtain for the Markov property are the following:

(1) If $f$ is an affine function, then $\{X_t\}$ is always a Markov field. This can be proved by a direct argument.

(2) In dimension one, assuming that $f$ and $g$ are twice continuously differentiable, $T$ is bijective, $g'(x) > -1$, $g'$ is not identically zero, and (i) and (ii') hold, then if $\{X_t\}$ is a Markov field, one necessarily has $f'' \equiv 0$.

(3) Suppose that

$$X_0^{i_k} = a_k; \qquad 1 \le k \le l$$
$$X_1^{j_k} = b_k; \qquad 1 \le k \le d - l$$

and $f$ is triangular, that means, $f^k(x)$ is a function of $x^1, \ldots, x^k$ for all $k$. In this case, if for each $k$, $f^k$ verifies a Lipschitz and linear growth condition on the variable $x^k$, one can show that there exists a unique solution of the equation $dX_t + f(X_t) = dW_t$ with the above boundary conditions, and the solution is a Markov process. This result is also proved by direct methods, and it says that unlike the one–dimensional case the solution of (4.4) can be a Markov process even though $f$ is non linear.

(4) In the following example in dimension 2 one also has a dicothomy similar to the one–dimensional case. Consider the equation

$$dX_t + f(X_t) = dW_t$$
$$X_1^1 = X_0^2 = 0$$

where $f(x^1, x^2) = (x^1 - x^2, -f_2(x^1))$ and $f_2$ is a twice continuously differentiable function such that $0 \leq f_2'(x) \leq K$ for some positive constant $K$. Then there exists a unique solution which is a Markov field only if $f_2'' \equiv 0$.

(5) In dimension one, and assuming a linear boundary condition of the type $F_0 X_0 + F_1 X_1 = h_0$, Donati–Martin (cf. [2]) has obtained the existence and uniqueness of a solution for the equation

$$dX_t = \sigma(X_t) \circ dW_t + b(X_t),$$

when the coefficients $b$ and $\sigma$ are of class $C^4$ with bounded derivatives, and $F_0 F_1 \neq 0$. On the other hand, if $\sigma$ is linear ($\sigma(x) = \alpha x$), $h_0 \neq 0$, and assuming that $b$ is of class $C^2$, then one can show that the solution $\{X_t\}$ is a Markov field only if the drift is of the form $b(x) = Ax + Bx \log|x|$, where $|B| < 1$.

For other types of existence results for first order stochastic differential equations with boundary conditions we refer to Dembo and Zeitouni [1] and Huang Zhiyuan [18].

## References

[1] Dembo, A., Zeitouni, O.: Existence of solutions for two–point stochastic boundary value problems. Preprint.

[2] Donati–Martin, C.: Equations différentielles stochastiques dans $\mathbb{R}$ avec conditions au bord. Preprint.

[3] Kusuoka, S.: The nonlinear transformation of Gaussian measure on Banach space and its absolute continuity (I). *J. Fac. Sci. Univ. Tokyo. Sec. IA*, 567–597 (1982).

[4] Mandrekar, V.: Germ–field Markov property for multiparameter processes, in *Seminaire de Probabilites X*. Lecture Notes in Math. **511**, 78–85 (1976)

[5] Nualart, D.: Noncausal stochastic integrals and calculus, in *Stochastic Analysis and Related Topics*, H. Korezlioglu & A.S. Ustunel Eds., Lecture Notes in Math. **1316**, 80-129 (1988).

[6] Nualart, D., Pardoux, E.: Stochastic calculus with anticipating integrands. *Probab. Theory Rel. Fields* **78**, 535-581 (1988).

[7] Nualart, D., Pardoux, E.: Boundary value problems for stochastic differential equations. Preprint.

[8] Nualart, D., Pardoux, E.: Second order stochastic differential equations with Dirichlet boundary conditions. Preprint.

[9] Nualart, D., Zakai, M.: Generalized stochastic integrals and the Malliavin Calculus. *Probab. Theory Rel. Fields* **73**, 255–280 (1986).

[10] Nualart, D., Zakai, M.: On the relation between the Stratonovich and Ogawa integrals. *Ann. Probab.* **17**, 1536-1540 (1989).

[11] Ocone, D., Pardoux, E.: Linear stochastic differential equations with boundary conditions. *Probab. Theory Rel. Fields* **82**, 489-526 (1989).

[12] Pardoux, E.: Applications of anticipating stochastic calculus to stochastic differential equations. Preprint.

[13] Ramer, R.: On non–linear transformations of Gaussian measures. *J. Funct. Anal.* **15**, 166–187 (1974).

[14] Russek, A.: Gaussian n–Markovian processes and stochastic boundary value problems. *Z. Wahrscheinlichkeittheorie und verw. Gebiete* **53**, 117–122 (1980).

[15] Simon, B.: *Trace ideals and their applications*, London Mathematical Society, Lecture Note Series 35, Cambridge University Press 1979.

[16] Skorohod, A.V.: On a generalization of a stochastic integral. *Theory Probab. Appl.* **20**, 219-233 (1975).

[17] Smithies, F.: *Integral Equations*. Cambridge Univ. Press, 1958.

[18] Huang Zhiyuan: On the generalized sample solutions of stochastic boundary value problems. *Stochastics* **11**, 237-248 (1984).

# SOBOLEV INEQUALITIES AND POLYNOMIAL DECAY OF CONVOLUTION POWERS AND RANDOM WALKS

by Laurent Saloff-Coste

Université de Paris VI - CNRS
Laboratoire Analyse Complexe et Géométrie
4, place Jussieu - Tour 46-0, 5ème étage
75752 PARIS CEDEX 05

# 1   Introduction.

Let $F$ be a symmetric, finitely supported, positive function on $\mathbf{Z}^d$ which charges the closest neighbors of the origin and has total mass one. Denote by $F^{(n)}$ the $n$-th convolution power of $F$. One has the following well known result:

$$\|F^{(n)}\|_\infty \le cn^{-d/2} , \quad n \in \mathbf{N}^* . \tag{1}$$

It is natural to ask what happens if one looks at more general Markov chains, namely the ones governed by a symmetric, bounded, Markov kernel $p(x, y), (x, y) \in (\mathbf{Z}^d)^2$, which charges the closest neighbors of $x$, uniformly on $\mathbf{Z}^d$. In this case, one can still hope that

$$p^{(n)}(x, y) \le cn^{-d/2} , \quad n \in \mathbf{N}^* \tag{2}$$

holds, where $p^n(x, y) = \int_{\mathbf{Z}^d} p(x, z)p^{(n-1)}(z, y)\mathrm{d}z$. However, the classical proof of (1) relies heavily on Fourier analysis and, as such, cannot be easily adapted to get a proof of (2). For the same reason, this classical approach fails to apply to more general groups or when there is no group structure at all.

This article reports on the work of the author [S] and a joint work with Th. Coulhon [C.S.1] where we extended some results of N. Varopoulos [V1], [V2], [V3], [V4], [V5], and simplified their proofs. The papers [C.K.S] and [C] contains also methods and results which are related to the ones we are going to describe.

Let $(X, dx)$ be a $\sigma$-finite measure space and $P$ be a symmetric submarkovian operator on the $L^p$ spaces over $(X, dx)$. Following the main ideas introduced by N. Varopoulos in [V1], [V2], we are going to characterize the possible polynomial decay of $\|P^k\|_{1\to\infty}$ in terms of certain Sobolev inequalities involving the Sobolev norms

$$\|(I - P)^{\alpha/2}f\|_2^2 = \langle(I - P)^\alpha f, f\rangle , \quad f \in L^2 , \quad \alpha > 0 .$$

More precisely, we have

**Theorem 1** *Let $P$ be as above and such that $\|P\|_{1\to\infty} < +\infty$. For any fixed $d > 0$ the following properties are equivalent.*

*i)* $\|P^k\|_{1\to\infty} \leq ck^{-d/2}$, $k \in \mathbb{N}^*$.

*ii)* $\|Pf\|_{2d/(d-2\alpha)} \leq c'\|(I-P)^{\alpha/2}f\|_2$, $f \in L^2$, *for some $\alpha$ such that* $0 < \alpha < d/2$.

*iii) For any $\alpha$ such that $0 < \alpha < d/2$, and any operator $S$ such that* $\|S\|_{p\to p} < +\infty$, $1 \leq p \leq +\infty$ *and* $\|S\|_{1\to\infty} < +\infty$, *one has*

$$\|Sf\|_{2d/(d-2\alpha)} \leq c_{\alpha,S}\|(I-P)^{\alpha/2}f\|_2, \quad f \in L^2.$$

Theorem 1 will be proved in section 2. The next result, which associated with (1) gives a proof of (2), is a corollary of Theorem 1.

**Theorem 2** *Let $p_i, i = 1,2$ be two symmetric Markov kernels on $(X, dx)$ such that*

$$\sup_{x,y\in X} \{p_i(x,y)\} < +\infty, \quad i = 1,2.$$

*Assume that there exist $c > 0$ and $d > 0$ such that*

$$p_1(x,y) \leq cp_2(x,y), \quad (x,y) \in X^2$$

*and*

$$\sup_{x,y\in X} \{p_1^{(k)}(x,y)\} \leq ck^{-d/2}, \quad k \in \mathbb{N}^*.$$

*Then, $p_2$ satisfies*

$$\sup_{x,y\in X} \{p_2^{(k)}(x,y)\} \leq c'k^{-d/2}, \quad k \in \mathbb{N}^*.$$

Here is how Theorem 2 can be deduced from Theorem 1. Let $P_i$ be the Markov operator associated with $p_i$ :

$$P_if(x) = \int_X p_i(x,y)f(y)dy, f \in L^1 + L^\infty,$$

and let us suppose first that $d > 2$. Using the assertion i) implies iii) of Theorem 1 with $P = P_1, \alpha = 1, S = P_2$, and the hypothesis of Theorem 2, we obtain

$$\|P_2f\|_{2d/(d-2)} \leq c\|(I-P)^{1/2}f\|_2 = c\langle(I-P_1)f,f\rangle^{1/2}. \tag{3}$$

Remark that for any symmetric Markov operator $P$ with kernel $p$, we have

$$\int_X\int_X |f(x) - f(y)|^2 p(x,y)dxdy = 2\langle(I-P)f,f\rangle. \tag{4}$$

This shows that the hypothesis $p_1(x,y) \leq cp_2(x,y)$ implies in particular that

$$\|(I-P_1)^{1/2}f\|_2^2 = \langle(I-P_1)f,f\rangle \leq c\langle(I-P_2)f,f\rangle = c\|(I-P_2)^{1/2}f\|_2^2. \tag{5}$$

From (3) and (5) we deduce at once that we have

$$\|P_2 f\|_{2d/(d-2)} \le c\|(I - P_2)^{1/2} f\|_2 \,, \quad f \in L^2$$

which implies, using Theorem 1 again, that

$$\|P_2^k\|_{1\to\infty} \le ck^{-d/2} \,, \quad k \in \mathbf{N}^* \,.$$

To make this proof work when $0 < d \le 2$, pick $\alpha_0$ such that $0 < \alpha_0 < d/2$ and remark that

$$\langle (I - P_1)f, f \rangle \le c\langle (I - P_2)f, f \rangle \,, \quad f \in L^2$$

implies, since $0 < \alpha_0 < d/2 \le 1$

$$\langle (I - P_1)^{\alpha_0} f, f \rangle \le c'\langle (I - P_2)^{\alpha_0} f, f \rangle \,, \quad f \in L^2$$

(for a proof of a more general fact see [D], p.110). Armed with this fact, one can rerun the preceding argument with $\alpha_0 = \alpha$ instead of $1 = \alpha$ and complete the proof of Theorem 2.

# 2　Decay of $\|P^k\|_{1\to\infty}$ and Sobolev inequalities.

Let us begin this section by an observation which turns out to be a powerful tool in the study of the behavior of $\|P^k\|_{1\to\infty}$. The idea behind the next proposition goes back to Hardy-Littlewood [H.L] and the simple proof given here and in [C.S.1] is adapted from [C].

**Proposition 1** *Let $P$ be a submarkovian symmetric operator and suppose that $\|P\|_{1\to\infty} < +\infty$. Then, for $d > 0$, the following properties are equivalent.*
*i)* $\|P^k\|_{1\to\infty} \le ck^{-d/2} \,, \quad k \in \mathbf{N}^*.$
*ii)* $\|P^k\|_{1\to\infty} \le c'k^{-d(1/p-1/q)/2} \,, \quad k \in \mathbf{N}^*,$ *for some $1 \le p < q \le +\infty$.*

**Proof.** Using classical interpolation results, we see that i) implies

$$\|P^k\|_{p\to q} \le ck^{-d(1/p-1/q)/2} \,, \quad k \in \mathbf{N}^* \,,$$

for any $1 \le p < q \le +\infty$. Now, let us assume that ii) is satisfied by $P$. We first claim that we have

$$\|P^k\|_{1\to q} \le c_q k^{-d(1-1/q)/2} = c_q k^{-d/2q'} \,, \quad k \in \mathbf{N}^*$$

where $1/q + 1/q' = 1$. Let $\theta \in ]0,1]$ be such that $1/p = \theta + (1-\theta)/q$ (i.e $1/p - 1/q = \theta/q'$). Indeed, by the hypothesis we have

$$\|P^{2n} f\|_q \le c_{p,q} n^{-d\theta/2q'} \|P^n f\|_p \,.$$

Also, by Hölder inequality, we have

$$\|P^n f\|_p \le \|P^n f\|_1^\theta \|P^n f\|_q^{1-\theta} \le \|f\|_1^\theta \|P^n f\|_q^{1-\theta} ,$$

and thus

$$\|P^{2n} f\|_q \le c_{p,q} n^{-d\theta/2q'} \|P^n f\|_q^{1-\theta} \|f\|_1^\theta .$$

Setting $K(N,f) = \sup_{n \in \{1,...,N\}} \{ n^{d/2q'} \|P^n f\|_q / \|f\|_1 \}$, we deduce at once from the preceding and $\|P\|_{1 \to q} < +\infty$ that

$$\|P^n f\|_q \le c n^{-d/2q'} K(N,f)^{1-\theta} \|f\|_1 , \quad \forall n \in \{1,...N\} ,$$

and finally, $K(N,f) \le c^{1/\theta}$, which proves the claim. By a duality argument, we obtain

$$\|P^k\|_{q' \to \infty} \le c k^{-d/2q'} , \quad k \in \mathbf{N}^* ,$$

which, by the same argument as above, gives i).

Our next goal is to prove the assertion "i) implies iii)" of Theorem 1. For $\alpha \in \mathbf{R}$, let $a_n(\alpha)$ be the coefficient of $x^n$ in the Taylor expansion of $(1-x)^{-\alpha/2}$ at the origin, and let us set formally

$$(I - P)^{-\alpha/2} = \sum_0^{+\infty} a_n(\alpha) P^n .$$

Remark that, for $\alpha \le 0$, $(I - P)^{-\alpha/2}$ is in fact a bounded operator on all $L^p$-spaces when (say) $P$ is submarkovian and symmetric. Moreover, classical estimates yield

$$|a_n(\alpha)| \le c_\alpha n^{\alpha/2-1} , \quad n \in \mathbf{N}^* .$$

This proves that, under the assumption that $\|P^k\|_{1 \to \infty} \le c k^{-d/2}, k \in \mathbf{N}^*$, $(I - P)^{-\alpha/2}$ sends $L^p$ into $L^p + L^\infty$ for all $\alpha > 0$ and $p \ge 1$ such that $\alpha p < d$. More precisely, we have

**Proposition 2** *Let $P$ be a submarkovian symmetric operator and suppose that $P$ satisfies $\|P^k\|_{1 \to \infty} \le c k^{-d/2}, k \in \mathbf{N}^*$. Then, for all $\alpha > 0, p > 1$ such that $\alpha p < d$, there exists a constant $c' > 0$ such that*

$$\left\| \left( (I - P)^{-\alpha/2} - I \right) f \right\|_{dp/(d-p\alpha)} \le c' \|f\|_p , \quad f \in L^p .$$

**Proof.** Let us denote by $f^*$ the maximal function $f^* = \sup_{n \ge 1} \{ |P^n f| \}$. By a theorem of E. Stein [S], one knows that there exists $c_p$, depending only on $1 < p < +\infty$ such that :

$$\|f^*\|_p \le c_p \|f\|_p .$$

For $N \geq 1$, let us write

$$
\left|\left((I-P)^{-\alpha/2} - I\right)f\right| = \left|\sum_{1}^{N-1} a_n(\alpha)P^n f + \sum_{N}^{+\infty} a_n(\alpha)P^n f\right|
$$

$$
\leq c_1 N^{\alpha/2} f^* + c_2 N^{\alpha/2 - d/2p}\|f\|_p
$$

where we have used the estimate for $|a_n(\alpha)|$ and the fact that

$$
\|P^k\|_{p \to \infty} \leq ck^{-d/2p} , \quad k \in \mathbb{N}^* ,
$$

which follows from the easy part of Proposition 1 and the hypothesis. Remark that, in the definition of $f^*$, we started from $n = 1$ instead of $n = 0$. From this, it follows that we have $\|f^*\|_\infty \leq c\|f\|_p$. Thus, we can optimize the preceding inequality by choosing $N \sim (c_2\|f\|_p/c_1 f^*)^{2p/d}$. This gives

$$
\left|\left((I-P)^{-\alpha/2} - I\right)f\right| \leq (c_1 f^*)^{1-\alpha p/d}(c_2\|f\|_p)^{\alpha p/d} .
$$

Setting $q = dp/(d-\alpha p)$ (i.e $1/q = 1/p - \alpha/d$), we obtain

$$
\left\|\left((I-P)^{-\alpha/2} - I\right)f\right\|_q \leq c_1^{1-\alpha p/d} c_2^{\alpha p/d}\|f\|_p^{\alpha p/d}\|f^*\|_p^{p/q}
$$

$$
\leq c_1^{1-\alpha p/d} c_2^{\alpha p/d} c_p^{p/q}\|f\|_p .
$$

This ends the proof of Proposition 2. It is clear that Proposition 2 implies the assertion "i) implies iii)" in Theorem 1: if $S$ is an operator such that $\|S\|_{p \to p} < +\infty$ for any $1 \leq p \leq +\infty$ and $\|S\|_{1 \to \infty} < +\infty$, then, under the assumption that i) holds and that $0 < \alpha < d/2$, we have

$$
\left\|S\left((I-P)^{-\alpha/2} - I\right)f\right\|_{2d/(d-2\alpha)} \leq c\|f\|_2
$$

and

$$
\|S(I-P)^{-\alpha/2}f\|_{2d/(d-2\alpha)} \leq c\|f\|_2 + \|Sf\|_{2d/(d-2\alpha)} \leq c'\|f\|_2 .
$$

Finally, using the obvious group property of the operators $(I-P)^\beta$, we get

$$
\|Sf\|_{2d/(d-2\alpha)} \leq c\|(I-P)^{\alpha/2}f\|_2 , \quad f \in L^2 .
$$

**Remark.** The fact that $P$ is submarkovian symmetric does not play an important role in Propositions 1 and 2. What is really needed here is $P$ being a power bounded operator on the $L^p$-spaces, $1 \leq p \leq +\infty$; see [C.S.1].

We are now going to prove the statement "ii) implies i)" in Theorem 1. For technical reasons, consider the operator $\tilde{P} = (I+P)/2$. It is clear that $\tilde{P}$ is submarkovian and symmetric whenever $P$ is. It is also clear that, if $P$ is submarkovian symmetric, $\tilde{P}$ satisfies $(\tilde{P}f, f) \geq 0$, $f \in L^2$. From the above and elementary spectral theory, we deduce that

$$
\|(I-\tilde{P})^\alpha \tilde{P}^n\|_{2 \to 2} \leq c_\alpha n^{-\alpha} , \quad n \in \mathbb{N}^* \tag{6}
$$

for any $\alpha > 0$.

**Lemma 1** *Let $P$ be a submarkovian, symmetric operator and assume that $\|P\|_{1 \to \infty} < +\infty$. Assume also that $P$ satisfies*

$$\|Pf\|_{2d/(d-2\alpha)} \leq c\|(I-P)^{\alpha/2}f\|_2 \,, \quad f \in L^2$$

*for some $0 < \alpha < d/2$. Then, we have*

$$\|(P\tilde{P})^n\|_{1 \to \infty} \leq cn^{-d/2} \,, \quad n \in \mathbf{N}^* \,.$$

**Proof.** Write

$$P\tilde{P}^n = P(I - \tilde{P})^{-\alpha/2}(I - \tilde{P})^{\alpha/2}\tilde{P}^n \,.$$

Using our main hypothesis, the fact that $(I - \tilde{P}) = (I - P)/2$, and (6), we get

$$\|P\tilde{P}^n f\|_q \leq \|(I - \tilde{P})^{\alpha/2}\tilde{P}^n f\|_2 \leq cn^{-\alpha/2}\|f\|_2 \,.$$

From the preceding we deduce at once that

$$\|(P\tilde{P})^n f\|_{2 \to q} \leq cn^{-d(1/2 - 1/q)/2} \,, \quad n \in \mathbf{N}^*$$

because $\|P\|_{2 \to 2} \leq 1$ and $1/2 - 1/q = \alpha/d$. Moreover, we have $\|P\tilde{P}\|_{1 \to \infty} < +\infty$. Thus, we can apply Proposition 1 to the operator $P\tilde{P}$ and conclude that

$$\|(P\tilde{P})^n\|_{1 \to \infty} \leq cn^{-d/2} \,, \quad n \in \mathbf{N}^* \,.$$

**Lemma 2** *Let $P$ be a submarkovian symmetric operator. Assume that $P$ satisfies $\|P\|_{1 \to \infty} < +\infty$. Then, the following properties are equivalent.*
*i)* $\|P^n\|_{1 \to \infty} \leq cn^{-d/2} \,, \quad n \in \mathbf{N}^*.$
*ii)* $\|(P\tilde{P})^n\|_{1 \to \infty} \leq c'n^{-d/2} \,, \quad n \in \mathbf{N}^*.$

**Proof.** The fact that i) implies ii) is easy and without interest. To prove that ii) implies i), remark that for $0 \leq f \in L^2$, we have

$$\|(P\tilde{P})^n f\|_2^2 = \langle (P\tilde{P})^{2n} f, f \rangle = (1/2)^{2n} \sum_{i=0}^{2n} \sum_{=0}^{2n} C_{2n}^i \langle P^{i+2n} f, f \rangle$$

$$\geq (1/2)^{2n} \sum_{j=0}^{n} \langle P^{2j+2n} f, f \rangle$$

because all the terms are positive. Moreover, for $0 \leq j \leq n$, we have

$$\langle P^{2j+2n} f, f \rangle = \|P^{j+n} f\|_2^2 \geq \|P^{2n} f\|_2^2$$

from which we obtain

$$\|(P\tilde{P})^n f\|_2^2 \geq (1/2)\|P^{2n} f\|_2^2 \,, \quad n \in \mathbf{N}^* \,.$$

From this we deduce that ii) implies i), using Proposition 1 again. With Lemmas 1 and 2, we end the proof of Theorem 1.

# 3    Sobolev inequalities and Markov chains on groups having polynomial growth.

Let $(G, dy)$ be a locally compact, compactly generated, unimodular group endowed with its Haar measure. Let $\Omega$ be a fixed compact, symmetric and generating (i.e $G = \bigcup_1^{+\infty} \Omega^n$) neighborhood of the neutral element $e$ of $G$. Let $|A|$ be the measure of $A \subset G$ and define the volume growth function of $G$ by

$$V(n) = |\Omega^n| , \quad n \in \mathbb{N} .$$

It is not hard to see that for two neighborhoods of $e$ as above, say $\Omega_1$ and $\Omega_2$, we have

$$c^{-1} V_1(c^{-1}n) \leq V_2(n) \leq c V_1(cn) , \quad n \in \mathbb{N} ,$$

where $c > 0$ is independent of $n$. In particular, we say that $G$ has polynomial volume growth of order $D$ if for one (or any) $\Omega$ as above there exists $c > 0$ such that : $c^{-1} \leq V(n)/n^D \leq c$.

Our aim in this section is to prove the following broad generalization of (2).

**Theorem 3** *Let $G$ be a group having polynomial volume growth of order $D > 0$ and $U$ be an open generating neighborhood of $e$ in $G$. Let $p(x, y)$ be a symmetric Markov kernel on $G$ such that $\sup_{x,y}\{p(x, y)\} < +\infty$ and $\inf\{p(x, y) , \ y^{-1}x \in U\} > 0$. Then there exists $c > 0$ such that*

$$\sup_{x,y}\{p^{(n)}(x, y)\} \leq c n^{-D/2}, n \in \mathbb{N}^* .$$

The main tools we are going to use in the proof of the above theorem are Theorem 1 and some sharp comparisons of Sobolev's norms which we want now to describe. Let us denote by $P_0$ the Markov operator associated with the right convolution by the function $F_0$ defined by $F_0(x) = |\Omega^3|^{-1}$ if $x \in \Omega^3$, and $F_0(x) = 0$ otherwise. Namely,

$$P_0 f(x) = \int_G f(y) F_0(y^{-1}x) dy = |\Omega^3|^{-1} \int_{y^{-1}x \in \Omega^3} f(y) dy , \ f \in L^1 + L^\infty .$$

Let us define $\rho(x) = \inf\{n/x \in \Omega^n\}$ and denote by $f_h$ the function $x \mapsto f_h(x) = f(xh)$ and by $_hf$ the function $x \mapsto _h f(x) = f(hx)$.

**Lemma 3** *There exists $c > 0$ such that*

$$\|f_h - f\|_2 \leq c\rho(h)\|(I - P_0)^{1/2}f\|_2$$

*for all $h \in G$ and all $f \in L^2$.*

**Proof.** Fix $h \in G$ such that $\rho(h) = n$. We can find $y_0 = e, y_1, \ldots, y_n = h$ such that $y_{i+1} \in y_i \Omega$, $i = 0, \ldots, n-1$ (i.e $\rho(y_i^{-1} y_{i+1}) \leq 1$). Moreover, for any $x_1, \ldots, x_{n-1}$, we have

$$|f(h) - f(e)|^2 \leq n \left( |f(x_1) - f(e)|^2 + \ldots |f(x_{n-1}) - f(h)|^2 \right)$$

and, integrating against the measure $dx_1 \times \ldots \times dx_{n-1}$ over the set $y_1 \Omega \times y_2 \Omega \times \ldots \times y_{n-1} \Omega$, we get

$$
\begin{aligned}
|f(h) - f(e)|^2 \leq \ &n \Bigg( |\Omega|^{-1} \int_{y_1 \Omega} |f(x_1) - f(e)|^2 dx_1 + \\
&|\Omega|^{-2} \int_{y_1 \Omega} \int_{y_2 \Omega} |f(x_2) - f(x_1)|^2 dx_1 dx_2 + \ldots + \\
&|\Omega|^{-1} \int_{y_{n-1} \Omega} |f(x_{n-1}) - f(h)|^2 dx_{n-1} \Bigg).
\end{aligned}
$$

Remark that

$$
\begin{aligned}
|\Omega|^{-1} \int_{y_1 \Omega} |f(x_1) - f(e)|^2 dx_1 &\leq |\Omega|^{-1} \int_{\Omega^3} |f(y) - f(e)|^2 dy \\
&= \int_G \int_G |f(zy) - f(z)|^2 F_0(y) dy d\mu_0(z)
\end{aligned}
$$

where $\mu_0 = |\Omega^3| |\Omega|^{-1} \delta_e$ ($\delta_e$ is the Dirac mass at the point $e \in G$). In the same way, we have

$$
\begin{aligned}
|\Omega|^{-1} \int_{y_{n-1} \Omega} |f(x_{n-1}) - f(h)|^2 dx_{n-1} \\
= |\Omega|^{-1} \int_{\Omega^3} |f(hy) - f(h)|^2 dy \\
= \int_G \int_G |f(zy) - f(z)|^2 F_0(y) dy d\mu_{n-1}(z)
\end{aligned}
$$

where $\mu_{n-1} = |\Omega^3| |\Omega|^{-1} \delta_h$, as well as

$$
\begin{aligned}
|\Omega|^{-2} \int_{y_i \Omega} \int_{y_{i+1} \Omega} |f(x_{i+1}) - f(x_i)|^2 dx_i dx_{i+1} \\
\leq |\Omega|^{-2} \int_{y_i \Omega} \int_{\Omega^3} |f(x_i y) - f(x_i)| dx_i dy \\
= \int_G \int_G |f(zy) - f(z)|^2 F_0(y) dy d\mu_i(z)
\end{aligned}
$$

where $d\mu_i(z) = |\Omega^3| |\Omega|^{-2} 1_{y_i \Omega}(z) dz$. Hence, we have

$$|f(h) - f(e)|^2 \leq cn \int_G \int_G |f(zy) - f(z)|^2 F_0(y) dy d\mu_h(z)$$

where $\mu_h = \sum_{i=0}^{n-1} \mu_i$ satisfies $\|\mu_h\| \le cn$ (here $\|\mu_h\|$ is the total mass of the measure $\mu_h$). Applying the above to the functions $_x f, x \in G$ and integrating with respect to $x$ against the Haar measure $dx$, we obtain

$$\|f_h - f\|_2^2 \le cn\|\mu_h\| \int_G \int_G |f(xy) - f(x)|^2 F_0(y) dy dx$$

$$\le c'\rho^2(h)\|(I - P_0)^{1/2} f\|_2^2 ,$$

since $\rho(h) = n$ and (see (4))

$$\int_G \int_G |f(xy) - f(x)|^2 F_0(y) dy dx = 2\|(I - P_0)^{1/2} f\|_2^2 .$$

**Lemma 4** *Let $G$ be a group having polynomial volume growth. For any $0 < \alpha < 1$ there exists $c_\alpha$ such that :*

$$\int_G \left( \|f_h - f\|_2 \rho^{-\alpha}(h) \right)^2 V(h)^{-1} dh \le c_\alpha \|(I - P_0)^{1/2} f\|_2^2 , \quad f \in L^2 .$$

**Proof.** Setting $A = I - P_0$ we introduce the symmetric Markov semigroup

$$T_t = e^{-tA} = e^{-t(I-P_0)} = e^{-t} \sum_0^{+\infty} (tP_0)^n / n! .$$

From elementary spectral theory, it follows that

$$\|A^\alpha T_t\|_{2\to 2} \le c_\alpha t^\alpha , \quad t > 0 , \quad \alpha > 0 \qquad (7)$$

and

$$\langle A^\alpha f, f \rangle = \|A^{\alpha/2} f\|_2^2 = c'_\alpha \int_0^{+\infty} \left( t^{1-\alpha/2} \|A T_t f\|_2 \right)^2 \frac{dt}{t} \qquad (8)$$

for $f \in L^2$ and $0 < \alpha < 1$. Moreover, with this notation the result of Lemma 3 becomes

$$\|f_h - f\|_2^2 \le c\rho^2(h) \|A^{1/2} f\|_2^2 , \quad f \in L^2 . \qquad (9)$$

For $x \in G, h \in G$ and $t > 0$, write

$$f(xh) - f(x) = \int_0^t (A T_s f(xh) - A T_s f(x)) \, ds + T_t f(xh) - T_t f(x) .$$

It follows that

$$\|f_h - f\|_2 \le 2 \int_0^t \|A T_s f\|_2 ds + \|(T_t f)_h - T_t f\|_2 .$$

By (9), we also have

$$\|(T_t f)_h - T_t f\|_2 \le c\rho(h) \|A^{1/2} T_t f\|_2$$

$$\le c'\rho(h) \int_{t/2}^{+\infty} s^{-1/2} \|A T_s f\|_2 ds$$

where we have used (7) to obtain the last inequality. Choosing $t = \rho^2(h)$, we deduce from the above that

$$\int_G \left( \|f_h - f\|_2 \rho^{-\alpha}(h) \right)^2 V(h)^{-1} dh \leq c \int_G \left( \int_0^{+\infty} (\rho(h)^{-\alpha} 1_{\{s \leq \rho^2(h)\}} \right.$$

$$\left. + \rho(h)^{1-\alpha} s^{-1/2} 1_{\{2s \geq \rho^2(h)\}} \right) \|AT_s f\|_2 ds \bigg)^2 \frac{dh}{V(h)}$$

The right hand side of this inequality is of the form $\int g^2(h) V(h)^{-1} dh$ with

$$g(h) = \int_0^{+\infty} K(h, s) s^{1-\alpha/2} \|AT_s f\|_2 \frac{ds}{s}$$

where we have set

$$K(h, s) = c \left( \rho(h)^{-\alpha} s^{\alpha/2} 1_{\{s \leq \rho^2(h)\}} + \rho(h)^{1-\alpha} s^{-(1-\alpha)/2} 1_{\{2s \geq \rho^2(h)\}} \right).$$

Easy calculations, using the hypothesis that $G$ has polynomial volume growth, show that $\int K(h, s) \frac{ds}{s} \leq c$ and $\int K(h, s) V(h)^{-1} dh \leq c$. These estimates on $K$ imply classically that the operator $T_K$ defined by

$$T_K \varphi(h) = \int_0^{+\infty} K(h, s) \varphi(s) \frac{ds}{s}$$

is bounded from $L^2 ([0, +\infty[, ds/s)$ to $L^2 (G, V(h)^{-1} dh)$. Finally, using (8), we obtain

$$\int_G \left( \|f_h - f\|_2 \rho^{-\alpha}(h) \right)^2 \frac{dh}{V(h)} \leq \int g^2(h) \frac{dh}{V(h)}$$

$$\leq c \int_0^{+\infty} \left( s^{1-\alpha/2} \|AT_s f\|_2 \right)^2 \frac{ds}{s} = c' \|(I - P_0)^{1/2} f\|_2^2 .$$

This end the proof of Lemma 4. Now, remark that the quantity on the left hand side of the inequality in Lemma 4 satisfies

$$\int_G \left( \|f_h - f\|_2 \rho^{-\alpha}(h) \right)^2 \frac{dh}{V(h)} \simeq \int \|f_h - f\|_2^2 F_\alpha(h) dh$$

where the function $F_\alpha$ is defined by :

$$F_\alpha = c_\alpha \sum_1^{+\infty} j^{-2\alpha-1} V(j)^{-1} 1_{\Omega^j}$$

with $c_\alpha^{-1} = \sum_1^{+\infty} j^{-2\alpha-1}$. Concerning the functions $F_\alpha$, we have :

**Lemma 5** Let $G$ be a unimodular locally compact group and let $(B_j)_{\mathbf{N}^*}$ be a sequence of subsets of $G$. Let $\gamma_j$ be the measure of $B_j$ and $\chi_j$ the characteristic function of $B_j$. Let $F_\alpha$ be the function defined by $F_\alpha = c_\alpha \sum_{i=1}^{+\infty} j^{-2\alpha-1} \gamma_j^{-1} \chi_j$, where $c_\alpha$ is such that $\int F_\alpha = 1$. Then, if $\gamma_j \geq c j^D$ for some $D > 0$ and all $j \geq 1$, one has

$$\|F_\alpha^{(n)}\|_\infty \leq c n^{-D/2\alpha} , \quad n \in \mathbf{N}^* .$$

**Proof.** Let $F = \sum_1 \lambda_j \varphi_j$ where $\int \varphi_j = 1$, $\|\varphi_j\|_\infty = \sigma_j^{-1}$, $\varphi_j > 0$, $\sum \lambda_j = 1$. Write

$$F^{(n)} = \left(\sum_1^\infty \lambda_j \varphi_j\right)^{(n)} = \sum_{(i_1,\dots,i_n)} \lambda_{i_1}\dots\lambda_{i_n}\varphi_{i_1} * \dots * \varphi_{i_n} .$$

From this, it follows that

$$\|F^{(n)}\|_\infty \leq \sum_{(i_1,\dots,i_n)} \lambda_{i_1}\dots\lambda_{i_n}\sigma_{\sup(i_1,\dots,i_n)}^{-1}$$

$$= \sum_k ((\lambda_1 + \dots + \lambda_k)^n - (\lambda_1 + \dots + \lambda_{k-1})^n)\,\sigma_k^{-1}$$

$$\leq n\sum_{k=1}^{+\infty} \lambda_k(\lambda_1 + \dots + \lambda_k)^{n-1}\sigma_k^{-1}$$

$$\leq n\sum_{k=1}^{+\infty} \lambda_k e^{(n-1)\log\left(1-\sum_{k+1}^\infty \lambda_j\right)}\sigma_k^{-1}$$

$$\leq n\sum_1^{+\infty} \lambda_k e^{-(n-1)\sum_{k+1}^\infty \lambda_j}\sigma_k^{-1}$$

Applying this to $F_\alpha$, we obtain

$$\|F_\alpha^{(n)}\|_\infty \leq c_\alpha n\sum_{k=1}^{+\infty} j^{-D-2\alpha-1}e^{-nc_\alpha' j^{-2\alpha}} .$$

It is not hard to see that the value of $c_\alpha \sum_1^{+\infty} j^{-D-2\alpha-1}e^{-nc_\alpha' j^{-2\alpha}}$, $n \in \mathbf{N}^*$, is comparable to the value of $c \int_0^{+\infty} t^{-D-2\alpha-1}e^{-nc_\alpha' t^{-2\alpha}}\,dt$ which is

$$c_\alpha \left(\int_0^{+\infty} t^{-D-2\alpha-1}e^{-c_\alpha' t^{-2\alpha}}\,dt\right) n^{-1-D/2\alpha} .$$

This yields

$$\|F_\alpha^{(n)}\|_\infty \leq c_\alpha n^{-D/2\alpha} , \quad n \in \mathbf{N}^* ,$$

and ends the proof of Lemma 5.

Armed with all the preceding results we can give a proof of Theorem 3. Let $G$ be of polynomial growth of order $D > 0$. By Lemma 5, the function $F_\alpha = c_\alpha \sum j^{-2\alpha-1}V(j)^{-1}1_{\Omega_j}$ satisfies

$$\|F_\alpha^{(n)}\|_\infty \leq c_\alpha n^{-D/2\alpha} , \quad n \in \mathbf{N}^* .$$

Theorem 1 tells us that for any operator $S$ such that $\|S\|_{p\to p} < +\infty$, $1 \leq p \leq +\infty$, and $\|S\|_{1\to\infty} < +\infty$, one has

$$\|Sf\|_{2(D/\alpha)/(D/\alpha-2)}^2 \leq c_\alpha\|(I - \xi_\alpha)^{1/2}f\|_2^2$$

$$= c_\alpha' \int \|f_h - f\|_2^2 F_\alpha(h)\,dh , \quad f \in L^2$$

where $\xi_\alpha$ is operator of right convolution with $F_\alpha$. Using Lemma 4, we deduce that we have

$$\|Sf\|^2_{2D/(D-2\alpha)} \le c_\alpha \|(I - P_0)^{\alpha/2}f\|^2_2, \quad f \in L^2,$$

which, using Theorem 1 again, yields

$$\|P^n_0\|_{1\to\infty} = \|F^{(n)}_0\|_\infty \le cn^{-D/2}, \quad n \in \mathbf{N}^*. \tag{10}$$

Now, let $p$ and $U$ be as in Theorem 3 and remark that the set of $x \in G$ such that there exists an integer $n_x$ for which $1^{(n_x)}_U(x) > 0$ is a subgroup of $G$ which contains $U$ and thus is $G$ itself. Remark also that if $1^{(n)}_U(x) > 0$ then $1^{(m)}_U(x) > 0$ for $m \ge n$. From this and the fact that $\Omega^3$ is compact, we deduce that there exists $n_0$ and $\delta > 0$ such that

$$p^{(n_0)}(x, y) \ge \epsilon^{n_0} 1^{(n_0)}_U(y^{-1}x) \ge \delta\epsilon^{n_0} 1_{\Omega^3}(y^{-1}x) \tag{11}$$

where $\epsilon = \inf\{p(x, y), y^{-1}x \in U\} > 0$. Finally, Theorem 2, (10) and (11) end the proof of Theorem 3.

# 4  Further results.

For groups which are not of polynomial volume growth, we still have the following result :

**Theorem 4** *Let $G$ be a locally compact, compactly generated unimodular group and $V$ be the volume growth function of $G$. Let $p(x, y)$ be a symmetric Markov kernel on $G$ and assume that $\sup_{x,y}\{p(x, y)\} < +\infty$ and $\inf\{p(x, y), y^{-1}x \in U\} > 0$, where $U$ is an open generating neighborhood of $e$ in $G$. If the function $V$ satisfies $V(n) \ge cn^A$, $n \in \mathbf{N}^*$, for some $A > 0$, then, for all $0 < a < A$, we have*

$$\sup_{x,y}\left\{p^{(n)}(x, y)\right\} \le c_a n^{-a/2}, \quad n \in \mathbf{N}^*.$$

**Remarks.** 1) The proof of the above theorem is along the same lines that the proof of Theorem 3, but it is simpler because it doesn't use Lemma 4 (which is restricted to group having polynomial growth); see [V5], [S], [C.S.1].

2) For a large class of groups, Theorems 3 and 4 take care of any possible case. This is true for solvable groups and for connected groups because, in both cases, one has the following result: either there exists $D \in \mathbf{N}$ such that $V(n) \simeq n^D$, $n \in \mathbf{N}^*$ or there exists $c > 0$ such that $V(n) \ge e^{cn}$, $n \in \mathbf{N}^*$. In the setting of finitely generated groups, either there exists $D \in \mathbf{N}$ such that $V(n) \simeq n^D$, $n \in \mathbf{N}^*$ or, for any $A \in \mathbf{N}$, there exists $c_A$ such that $V(n) \ge c_A n^A$, $n \in \mathbf{N}$. For more details on these deep and difficult results,

see [G], [Gr], [V4].

3) In a recent joint work with Th. Coulhon, we generalized Theorems 3 and 4 to some non symmetric Markov chains, namely those governed by a kernel satisfying $\int p(x, y)dx = \int p(x, y)dy = 1$. We also generalized theorem 2 to the case where $p_2$ satisfies the above condition instead of being symmetric. See [C.S.2].

4) An alternative way of getting the results presented above is to use Nash's inequalities instead of Sobolev's ones. See [C.K.S] and also [C.S.1], [C.S.2].

## References

[C.K.S] **E. Carlen, S. Kusuoka, D. Stroock.** Upper bounds for symmetric Markov transition functions. **Ann. Inst. H. Poincaré**, probabilités et statistiques, suppl. au #2, 1987, pp.245-287.

[C] **Th. Coulhon.** Dimension á l'infini d'un semi-groupe analytique. **Bull. Sci. Math.** 1990.

[C.S.1] **Th. Coupon, L. Saloff-Coste.** Puissances d'un opérateur régularisant. **Ann. Inst. H. Poincaré**, probabilités et statistiques, 26, 1990, pp.419-436.

[C.S.2] **Th. Coupon, L. Saloff-Coste.** Marches aléatoires non symétriques sur les groupes unimodulaires. **C. R. Acad. Sci. Paris**, 310, 1990, pp.627-630.

[D] **E.B. Davies.** One parameter semi-groups. Academic press, 1980.

[Gr] **M. Gromov.** Groups of polynomial growth and expanding maps. **Publ. Math. IHES**, 53, 1981, pp.53-78.

[G] **Y. Guivarc'h.** Croissance polynomiale et périodes des fonctions harmoniques. **Bull. Soc. Math. France**, 101, 1973, pp.333-379.

[H.L] **G. Hardy, J. Littlewood.** Some properties of conjugate functions. **J. Reine-Angew Math.** 167, 1932, pp.405-423.

[S] **L. Saloff-Coste.** Sur la décroissance des puissances de convolution sur les groupes. **Bull. Sci. Math.** 2ème série, 113, 1989, pp.3-21.

[St] **E. Stein.** Singular integrals and differentiability properties of functions. Princeton University Press, 1970.

[V1] **N. Varopoulos.** Chaînes de Markov et inégalités isopéri-
métriques. **C.R. Acad. Sci. Paris**, t.298, série I, 1984,
pp.233-236.

[V2] **N. Varopoulos.** Chaînes de Markov et inégalités isopéri-
métriques. **C.R. Acad. Sci. Paris**, t.298, série I, 1984,
pp.465-468.

[V3] **N. Varopoulos.** Isoperimetric inequalities and Markov
chains. **J. Funct. Anal.**, vol.63, 1985, pp.215-239.

[V4] **N. Varopoulos.** Théorie du potentiel sur des groupes et
des variétés. **C.R. Acad. Sci. Paris**, t.302, série I, 1986,
pp.203-205.

[V5] **N. Varopoulos.** Convolution powers on locally compact
groups. **Bull. Sc. Math.** 2ème série, 111, 1987, pp.333-
342.

# MICROCANONICAL DISTRIBUTIONS FOR ONE DIMENSIONAL LATTICE GASES

DANIEL W. STROOCK

January 1990

## §1: Gibbs' States

In equilibrium statistical mechanics, a central rôle is played by the **Gibbs' state** of a system, and the goal of this note is to understand, on the basis of large deviation theory, why Gibbs' states arise. In order to keep everything very simple, we will restrict our attention here to a very special class of systems known as **one dimensional lattice gases with finite range interaction** (cf. the Remark 2.15 below). To be precise, let $E$ be a compact metric space, $\lambda$ a probability measure on $(E, \mathcal{B})$, set $\Omega = E^{\mathbb{Z}}$, and let $\mathcal{U} = \{U_F : \emptyset \neq F \subset\subset \mathbb{Z}\} \subseteq C(\Omega; \mathbb{R})$ be a family of continuous functions with the properties that, for all non-empty, finite subsets $F$ of $\mathbb{Z}$,

**1)** $U_F(\mathbf{x}) = U_F(\mathbf{y})$ when $\mathbf{x}, \mathbf{y} \in \Omega$ with $x_k = y_k$ for all $k \notin F$;

**2)** for all $k \in \mathbb{Z}$, $U_{k+F} = U_F \circ S^k$, where $S : \Omega \longrightarrow \Omega$ is the **shift transformation** determined by $(S\mathbf{x})_k = x_{k+1}$ for all $k \in \mathbb{Z}$ and $\mathbf{x} \in \Omega$;

**3)** there is an $R \in \mathbb{Z}^+$ such that $U_F \equiv 0$ if $0 \in F$ and $F \not\subseteq [-R, R]$.

Such a family $\mathcal{U}$ is said to be a **shift-invariant potential with range** $R$.

Starting from $\mathcal{U}$, we define a the probability measures $\gamma_{\beta,\Lambda}(\cdot \,|\mathbf{y})$ on $(\Omega, \mathcal{B}_\Omega)$ for $\beta \in \mathbb{R}$, $\emptyset \neq \Lambda \subset\subset \mathbb{Z}$ and $\mathbf{y} \in \Omega$, so that, for $f \in C(\Omega; \mathbb{R})$,

$$\int_\Omega f(\mathbf{x})\, \gamma_{\beta,\Lambda}(d\mathbf{x}|\mathbf{y})$$

$$= \frac{1}{Z_\Lambda(\beta, \mathbf{y})} \int_\Omega f(\mathbf{x}_\Lambda \bullet \mathbf{y}_{\Lambda\bullet}) \exp\left[-\beta \sum_{F \cap \Lambda \neq \emptyset} |F|\, U_F(\mathbf{x}_\Lambda \bullet \mathbf{y}_{\Lambda\bullet})\right] \lambda^{\mathbb{Z}}(d\mathbf{x}),$$

where, for $\mathbf{x} \in \Omega$, $\mathbf{x}_\Lambda$ is the element of $E^\Lambda$ obtained by restricting $\mathbf{x}$ to $\Lambda$ and $\mathbf{x}_\Lambda \bullet \mathbf{y}_{\Lambda\bullet}$ is the element of $\Omega$ whose restrictions to $\Lambda$ and $\Lambda^\complement$ coincide with those of $\mathbf{x}$ and $\mathbf{y}$, respectively. (The number $Z_\Lambda(\beta, \mathbf{y})$ is determined

The author acknowledges support the grants NSF DMS–8611487 and DAAL 03–86–K–0171.

by the condition that $\gamma_{\beta,\Lambda}(\Omega|\mathbf{y}) = 1$.) It is then an easy matter to check that $\{\gamma_{\beta,\Lambda}(\cdot|\mathbf{y}) : \emptyset \neq \Lambda \subset\subset \mathbb{Z} \text{ and } \mathbf{y} \in \Omega\}$ is a consistent family of regular conditional probabilities in the sense that, for $\Lambda_1 \subset \Lambda_2$,

$$\int_\Omega f(\mathbf{x})\gamma_{\beta,\Lambda_2}(d\mathbf{x}|\mathbf{y})$$

$$= \int_\Omega \left( \int_\Omega f(\mathbf{x}_{\Lambda_1} \bullet \boldsymbol{\xi}_{\Lambda_1^\mathfrak{c}}) \, \gamma_{\beta,\Lambda_1}(d\mathbf{x}|\boldsymbol{\xi}_{\Lambda_2} \bullet \mathbf{y}_{\Lambda_2^\mathfrak{c}}) \right) \gamma_{\beta,\Lambda_2}(d\boldsymbol{\xi}|\mathbf{y}).$$

In particular, if $\mathbf{y}$ is any fixed element of $\Omega$ and $\gamma$ is any (weak) limit of the sequence $\{\gamma_{\beta,[-n,n]}(\cdot|\mathbf{y})\}_{n=0}^\infty$, then

(1.3) $$\int_\Omega f(\mathbf{x})\,\gamma(d\mathbf{x}) = \int_\Omega \left( \int_\Omega f(\mathbf{x}_\Lambda \bullet \mathbf{y}_{\Lambda^\mathfrak{c}}) \, \gamma_{\beta,\Lambda}(d\mathbf{x}|\mathbf{y}) \right) \gamma(d\mathbf{y})$$

for every $\emptyset \neq \Lambda \subset\subset \mathbb{Z}$. Moreover, under the stated conditions (in particular, because we are working with the lattice $\mathbb{Z}$ and not one of higher dimension), one can show that, for every $\beta \in \mathbb{R}$, there is only one $\gamma$ for which (1.3) can hold for all $\Lambda$'s, and we will call this unique $\gamma$ the **Gibbs' state for the potential** $\mathcal{U}$ **at reciprocal temperature** $\beta$ and will use $\gamma_\beta$ to denote it.

Gibbs' states turn up in statistical mechanics because they are supposed to be *the equilibrium distribution* of the system under consideration, and the reasoning which underlies this supposition is based on the following picture. Think of $E$ as being the *phase* space of an individual particle and of $\lambda$ as the *Liouville measure* for the dynamics of each particle when it is *free* (i.e., there is no *interaction*). Next, suppose that we place free particles in a line along the lattice $\mathbb{Z}$ and have them interact in such a way that the energy produced by the interaction of the particle at $k$ with the rest of the system is given by

$$\mathcal{U}_k(\mathbf{x}) \equiv \sum_{F \ni k} U_F(\mathbf{x}) = \mathcal{U}_0 \circ S^k$$

when the position (in $\Omega$) of the particles is $\mathbf{x}$. Finally, consider what happens when we allow our interacting system to achieve equilibrium subject only to the constraint that the *average interaction energy* of the particles be some specified number $\overline{U}$. To be more precise, let $n \in \mathbb{Z}^+$ be given and consider the system in which only the particles at sites $k \in [-n,n]$ interact but all the other particles are free. When such a system has achieved equilibrium subject only to the constraint that its average interaction energy be $\overline{U}_n$, one suspects that its distribution should be the measure $\mu_n$ which one gets by conditioning $\lambda^{\mathbb{Z}}$ on the event

$$A_n \equiv \left\{ \mathbf{x} \in \Omega : \frac{1}{2n+1} \sum_{|k| \leq n} \mathcal{U}_k(\mathbf{x}) = \overline{U} \right\}.$$

In the language of statistical mechanics, $\mu_n$ would be called the **microcani-cal distribution** of this system and what the **principle of equivalence of ensembles** predicts is that, as $n \to \infty$, $\mu_n$ tends to $\gamma_\beta$, where $\beta$ (the reciprocal temperature) is determined by the condition that

$$(1.4) \qquad \int_\Omega \mathcal{U}_0(\mathbf{x}) \gamma_\beta(dx) = \overline{U}.$$

The purpose of this note is to verify the equivalence of ensembles as an application of the theory of large deviations (cf. Theorem 2.15 below). Earlier programs of this sort have been carried out by Dobrushin and Tirozzi in the article [DT] and by Georgii in the book [G]: in [DT] the reasoning is based on the Central Limit Theorem whereas the ideas in [G] derive from de Finetti's theory of symmetric random variables. Thus, at best, all that is being proposed here is a new strategy for handling this sort of question. In fact, the strategy itself is not entirely new, since it has been used already to handle a closely related situation in [SZ].

## §2: The Equivalence of Ensembles

In order not to get involved with problems about the existence of regular conditional probability distributions, we will replace the true microcanonical distribution $\mu_n$ by the **approximate microcanical distribution** $\mu_{n,\delta}$, $\delta \in (0,1]$, which is the conditional distribution of $\lambda^{\mathbb{Z}}$ given the event

$$(2.1) \qquad A_n(\delta) \equiv \left\{ \mathbf{x} \in \Omega : \left| \frac{1}{2n+1} \sum_{|k| \leq n} \mathcal{U}_k(\mathbf{x}) - \overline{U} \right| \leq \delta \right\},$$

and only at the end will we pass to the limit as $\delta \searrow 0$.

Our analysis rests on two observations. The first of these is the characterization of $\gamma_\beta$ as the solution to an extremal problem. Namely, let $\mathbf{M}_1(\Omega)$ denote the space of all probability measures $\nu$ on $(\Omega, \mathcal{B}_\Omega)$ and let $\mathbf{M}_1^{\mathbf{S}}(\Omega)$ be the subset of $\nu \in \mathbf{M}_1(\Omega)$ which are shift-invariant (i.e., $\nu = \nu \circ S^{-1}$.) Notice that, as a consequence of uniqueness, not only is $\gamma_\beta$ an element of $\mathbf{M}_1^{\mathbf{S}}(\Omega)$ for every $\beta \in \mathbb{R}$, it is an *ergodic* one. Next, for any $\emptyset \neq \Lambda \subset\subset \mathbb{Z}$ and any pair of probability measures $\mu$ and $\nu$ on $(E^\Lambda, \mathcal{B}_{E^\Lambda})$, define the **entropy of $\mu$ relative to $\nu$** to be the number $\mathbf{H}(\mu|\nu)$ given by

$$\mathbf{H}(\mu|\nu) = \begin{cases} \int_{E^\Lambda} f \log f \, d\nu & \text{if } \nu << \mu \text{ and } f = \frac{d\mu}{d\nu} \\ \infty & \text{otherwise.} \end{cases}$$

If, for $\mu \in \mathbf{M}_1(\Omega)$ and non-empty $\Lambda \subseteq \mathbb{Z}$, we use $\mu_\Lambda$ to denote the marginal distribution of $\nu$ on $E^\Lambda$ (i.e.,

$$\int_{E^\Lambda} f \, d\mu_\Lambda = \int_\Omega f(\mathbf{x}_\Lambda) \, \mu(dx)$$

for $f \in C(E^{\Lambda}; \mathbb{R}))$, then (cf. Lemma 4.4.7 in [DS]) for disjoint $\Lambda_1$ and $\Lambda_2$ one has that

$$(2.2) \qquad \mathbf{H}\big(\mu_{\Lambda_1 \cup \Lambda_2} \big| \nu_1 \times \nu_2\big) \geq \mathbf{H}\big(\mu_{\Lambda_1} \big| \nu_1\big) + \mathbf{H}\big(\mu_{\Lambda_2} \big| \nu_2\big)$$

for any $\mu \in \mathbf{M}_1(\Omega)$ and probability measures $\nu_i$ on $\big(E^{\Lambda_i}, \mathcal{B}_{E^{\Lambda_i}}\big)$. Hence, if we set

$$\mathbf{H}_n(\nu) = \mathbf{H}\big(\nu_{[-n,n]} \big| \lambda^{[-n,n]}\big) \quad \text{for} \quad n \in \mathbb{Z}^+ \text{ and } \nu \in \mathbf{M}_1(\Omega),$$

one can then easily show (via subaddativity considerations) that the **specific entropy**

$$(2.3) \qquad\qquad \mathbf{H}(\nu) \equiv \lim_{n \to \infty} \frac{\mathbf{H}_n(\nu)}{2n+1}$$

exists for every $\nu \in \mathbf{M}_1^{\mathbf{S}}(\Omega)$. In fact, one finds that,

$$(2.4) \quad (2n+1)\mathbf{H}(\nu) \geq \mathbf{H}_n(\nu) \geq 0 \quad \text{for all} \quad n \in \mathbb{Z}^+ \text{ and } \nu \in \mathbf{M}_1^{\mathbf{S}}(\Omega).$$

We extend the definition of $\mathbf{H}$ to the whole of $\mathbf{M}_1(\Omega)$ by taking it to be identically infinite on $\mathbf{M}_1(\Omega) \setminus \mathbf{M}_1^{\mathbf{S}}(\Omega)$. With these preparations, we can now give our extremal characterization of $\gamma_\beta$.

**Notation:** In what follows, we will use the notation $\langle f, \nu \rangle$ to denote the integral of a function $f$ with respect to a measure $\nu$. In addition, given $\beta \in \mathbb{R}$ and $\emptyset \neq \Lambda \subset\subset \mathbb{Z}$, set

$$\mathcal{U}_\Lambda = \sum_{\emptyset \neq F \subseteq \Lambda} |F| U_F \quad \text{and} \quad Z_\beta(\Lambda) = \int_\Omega \exp\big[-\beta \mathcal{U}_\Lambda(\mathbf{x})\big] \lambda^{\mathbb{Z}}(d\mathbf{x})$$

and define $\gamma_{\beta,\Lambda}$ on $\big(E^{\Lambda}, \mathcal{B}_{E^{\Lambda}}\big)$ so that

$$\int_{E^{\Lambda}} f \, d\gamma_{\beta,\Lambda} = \frac{1}{Z_\beta(\Lambda)} \int_\Omega f(\mathbf{x}_\Lambda) \exp\big[-\beta \mathcal{U}_\Lambda(\mathbf{x})\big] \lambda^{\mathbb{Z}}(d\mathbf{x}), \qquad f \in C(E^{\Lambda}; \mathbb{R}).$$

Notice that, in general, $\gamma_{\beta,\Lambda} \neq \big(\gamma_\beta\big)_\Lambda$.

**Lemma 2.5.** *For each $\beta \in \mathbb{R}$ and $\mu \in \mathbf{M}_1^{\mathbf{S}}(\Omega)$,*

$$(2.6) \qquad \mathbf{H}(\mu) \geq \mathbf{H}(\gamma_\beta) + \beta\Big(\big\langle \mathcal{U}_0, \gamma_\beta \big\rangle - \big\langle \mathcal{U}_0, \mu \big\rangle\Big).$$

*Moreover, if $\langle \mathcal{U}_0, \mu \rangle = \langle \mathcal{U}_0, \gamma_\beta \rangle$ and $\mathbf{H}(\mu) = \mathbf{H}(\gamma_\beta)$, then $\mu = \gamma_\beta$.*

*Proof.* First note that

$$\mathbf{H}_n(\mu) = \mathbf{H}\big(\mu_{[-n,n]} \big| \gamma_{\beta,[-n,n]}\big) + \int_{E^{[-n,n]}} \log \frac{d\gamma_{\beta,[-n,n]}}{d\lambda^{[-n,n]}} \, d\mu_{[-n,n]}$$

$$= \mathbf{H}\big(\mu_{[-n,n]} \big| \gamma_{\beta,[-n,n]}\big) - \log Z_\beta\big([-n,n]\big) - \beta\big\langle \mathcal{U}_{[-n,n]}, \mu \big\rangle.$$

Next, observe that

$$\left| \mathcal{U}_{[-n,n]} - \sum_{F \cap [-n,n] \neq \emptyset} |F| U_F \right| \vee \left| \mathcal{U}_{[-n,n]} - \sum_{k=-n}^{n} \mathcal{U}_k \right| \leq M$$

for some $M < \infty$ and all $n \in \mathbb{Z}^+$. Hence, by taking $\mu = \gamma_\beta$, we see first that

$$\mathbf{H}(\gamma_\beta) = \lim_{n \to \infty} \frac{\log Z_\beta([-n,n])}{2n+1} - \beta \langle \mathcal{U}_0, \gamma_\beta \rangle;$$

and, having obtained this result, we then get, for general $\mu \in \mathbf{M}_1^S(\Omega)$, that

$$\mathbf{H}(\mu) = \lim_{n \to \infty} \frac{\mathbf{H}\big(\mu_{[-n,n]} \big| \gamma_{\beta,[-n,n]}\big)}{2n+1} + \mathbf{H}(\gamma_\beta) + \beta\Big(\langle \mathcal{U}_0, \gamma_\beta \rangle - \langle \mathcal{U}_0, \mu \rangle\Big).$$

Thus, (2.6) certainly holds. In fact, we now see that the rest of the lemma will be proved as soon as we show that

$$(2.7) \qquad \lim_{n \to \infty} \frac{\mathbf{H}\big(\mu_{[-n,n]} \big| \gamma_{\beta,[-n,n]}\big)}{2n+1} = 0 \implies \mu = \gamma_\beta.$$

In order to prove (2.7), let $d \geq R$ be given, set $I = \{0, \ldots, 2d-1\}$, and let $\Lambda_n = [1, 2nd] = \bigcup_{m=1}^{n}(m+I)$ for $n \in \mathbb{Z}^+$. Then, just as before, we can find an $M < \infty$ such that

$$\mathbf{H}\Big(\mu_{\Lambda_n} \Big| \gamma_{\beta,\Lambda_n}\Big) \geq -n|\beta|M + \mathbf{H}\left(\mu_{\Lambda_n} \Big| \prod_{m=1}^{n} \gamma_{\beta,m+I}\right)$$

$$\geq -n|\beta M| + n\mathbf{H}(\mu_I | \gamma_{\beta,I}) \geq -(n+1)|\beta|M + n\mathbf{H}\Big(\mu_I \Big| (\gamma_\beta)_I\Big),$$

where, in the passage to the second line, we used (2.2). In particular, we now see that

$$\mathbf{H}\Big(\mu_{[-nd,n]} \Big| \gamma_{\beta,[-nd,nd]}\Big) \geq -(n+1)|\beta|M + n\mathbf{H}\Big(\mu_{[-d,d]} \Big| (\gamma_\beta)_{[-d,d]}\Big),$$

and therefore that the hypothesis in (2.7) implies

$$\sup_{d \in \mathbb{Z}^+} \mathbf{H}\Big(\mu_{[-d,d]} \Big| (\gamma_\beta)_{[-d,d]}\Big) < \infty.$$

But this means that $\mu << \gamma_\beta$; and, because both $\mu$ and $\gamma_\beta$ are shift-invariant and $\gamma_\beta$ is ergodic, it follows that $\mu = \gamma_\beta$. $\square$

The second observation which we will need is that all of our consider-ations can be transferred to questions about the sequence of **empirical distribution functionals**

$$\mathbf{x} \in \Omega \longmapsto \mathbf{R}_n(\mathbf{x}) \equiv \frac{1}{2n+1} \sum_{k=-n}^{n} \delta_{S^k \mathbf{x}} \in \mathbf{M}_1(\Omega).$$

(Throughout, we take the topology on $\mathbf{M}_1(\Omega)$ to be the topology of weak convergence. Thus, $\mathbf{M}_1(\Omega)$ is itself of compact metric space and measur-ability means with respect to the corresponding Borel field.) Indeed, it is clear that an equally good description of the set $A_n(\delta)$ in (2.1) is as the set of $\mathbf{x} \in \Omega$ such that

$$(2.8) \qquad \mathbf{R}_n(\mathbf{x}) \in \mathfrak{M}(\overline{U}; \delta) \equiv \left\{ \nu \in \mathbf{M}_1(\Omega) : \left| \langle \mathcal{U}_0, \nu \rangle - \overline{U} \right| \le \delta \right\}.$$

In addition, since

$$\int_\Omega f \, d\lambda^{\mathbb{Z}} = \int_\Omega \langle f, \mathbf{R}_n(\mathbf{x}) \rangle \, \lambda^{\mathbb{Z}}(d\mathbf{x}), \quad n \in \mathbb{Z}^+,$$

for every bounded measurable $f : \Omega \longrightarrow \mathbb{R}$, we will have reached our goal once we show that

$$\varlimsup_{\delta \searrow 0} \varlimsup_{n \to \infty} \lambda^{\mathbb{Z}} \left( \mathbf{R}_n \notin G \,\middle|\, \mathbf{R}_n \in \mathfrak{M}(\overline{U}; \delta) \right) = 0$$

for every open neighborhood $G$ of $\gamma_\beta$.

With these preliminaries, we can now state and prove our result.

**Theorem 2.9.** *Set*

$$\mathfrak{M}(\overline{U}) = \left\{ \mu \in \mathbf{M}_1^S(\Omega) : \langle \mathcal{U}_0, \mu \rangle = \overline{U} \right\}$$

*and assume that*

$$(2.10) \qquad m(\overline{U}) \equiv \inf \left\{ \mathbf{H}(\mu) : \mu \in \mathfrak{M}(\overline{U}) \right\} < \infty.$$

*(Implicit in (2.10) is the assumption that $\mathfrak{M}(\overline{U}) \ne \emptyset$.) Then, for each $\delta \in (0, 1)$ there is an $N_\delta \in \mathbb{Z}^+$ such that (cf. (2.1))*

$$\lambda^{\mathbb{Z}} \big( A_n(\delta) \big) \ge \exp \left[ -\frac{m(\overline{U})}{1-\delta} \right] \quad \text{for} \quad n \ge N_\delta.$$

*In fact, for any measurable subset $A$ of $\mathbf{M}_1(\Omega)$,*

$$- \inf_{\nu \in A^\circ} \mathbf{I}_{\mathcal{U}}(\nu) \le \varliminf_{\delta \searrow 0} \varliminf_{n \to \infty} \frac{1}{2n+1} \log \left[ \lambda^{\mathbb{Z}} \big( \mathbf{R}_n \in A \,\big|\, A_n(\delta) \big) \right]$$

$$(2.11) \qquad \le \varlimsup_{\delta \searrow 0} \varlimsup_{n \to \infty} \frac{1}{2n+1} \log \left[ \lambda^{\mathbb{Z}} \big( \mathbf{R}_n \in A \,\big|\, A_n(\delta) \big) \right] - \le \inf_{\nu \in \overline{A}} \mathbf{I}_{\mathcal{U}}(\nu),$$

*where*

$$\mathbf{I}_{\mathcal{U}}(\nu) \equiv \begin{cases} \mathbf{H}(\nu) - m(\overline{U}) & \text{if } \nu \in \mathbf{M}_1^S(\Omega) \text{ and } \langle \mathcal{U}_0, \nu \rangle = \overline{U} \\ \infty & \text{otherwise.} \end{cases}$$

*In particular, if $G$ is an open neighborhood of the set*

$$\mathfrak{G}(\overline{U}) \equiv \left\{ \mu \in \mathfrak{M}(\overline{U}) : \mathbf{H}(\mu) = m(\overline{U}) \right\},$$

*then*

(2.12) $\qquad \overline{\lim_{\delta \searrow 0}} \ \overline{\lim_{n \to \infty}} \ \dfrac{1}{n} \log \left[ \lambda^{\mathbb{Z}} \left( \mathbf{R}_n \notin G \, \middle| \, A_n(\delta) \right) \right] \leq - \inf_{\nu \notin G} \mathbf{I}_{\mathcal{U}}(\nu) < 0.$

*Finally, when $\langle \mathcal{U}_0, \gamma_\beta \rangle = \overline{U}$ for some $\beta \in \mathbb{R}$, then $\gamma_\beta$ is the one and only element of $\mathfrak{G}(\overline{U})$, and therefore*

(2.13) $\qquad \lim_{\delta \searrow 0} \ \lim_{n \to \infty} \ \dfrac{\int_{A_n(\delta)} f \, d\lambda^{\mathbb{Z}}}{\lambda^{\mathbb{Z}}(A_n(\delta))} = \int_\Omega f \, d\gamma_\beta$

*for every $f \in C(\Omega; \mathbb{R})$.*

*Proof.* In view of Lemma 2.5, it is clear that $\mathfrak{G}(\overline{U}) = \{\gamma_\beta\}$ if $\gamma_\beta \in \mathfrak{M}(\overline{U})$. Hence, the final assertion will be proved once we prove (2.12); and since (2.12) is itself an immediate consequence of (2.11), all that remains is the derivation of (2.11).

Actually, (2.11) is an easy application of the *large deviation principle*

(2.14)
$$- \inf_{\nu \in A^\circ} \mathbf{H}(\nu) \leq \varliminf_{n \to \infty} \frac{1}{2n+1} \log \left[ \lambda^{\mathbb{Z}} (\mathbf{R}_n \in A) \right]$$
$$\leq \varlimsup_{n \to \infty} \frac{1}{2n+1} \log \left[ \lambda^{\mathbb{Z}} (\mathbf{R}_n \in A) \right] \leq - \inf_{\nu \in \overline{A}} \mathbf{H}(\nu)$$

for all $A \in \mathcal{B}_{\mathbf{M}_1(\Omega)}$. The statement in (2.14) is the special case of Donsker and Varadhan's theory of large deviations for the (process level) empirical distribution of a Markov chain (the chain in our case being $\lambda^{\mathbb{Z}}$), and two different proofs can be found in Sections IV.4 and V.4 of [DS].

Given (2.14), ones sees immediately that, for every $\delta \in (0,1)$ and $A \in \mathcal{B}_{\mathbf{M}_1(\Omega)}$,

$$- \inf \left\{ \mathbf{H}(\nu) : \nu \in A^\circ \cap \mathfrak{M}(\overline{U}) \right\}$$
$$\leq \varliminf_{n \to \infty} \frac{1}{2n+1} \log \left[ \lambda^{\mathbb{Z}} \left( \mathbf{R}_n \in A \cap \mathfrak{M}(\overline{U}; \delta) \right) \right]$$
$$\leq \varlimsup_{n \to \infty} \frac{1}{2n+1} \log \left[ \lambda^{\mathbb{Z}} \left( \mathbf{R}_n \in A \cap \mathfrak{M}(\overline{U}; \delta) \right) \right]$$
$$\leq - \inf \left\{ \mathbf{H}(\nu) : \nu \in \overline{A} \cap \mathfrak{M}(\overline{U}; \delta) \right\};$$

and clearly (2.11) follows immediately from this, (2.8), and the easily verified fact that

$$\inf \left\{ \mathbf{H}(\nu) : \nu \in \overline{A} \cap \mathfrak{M}(\overline{U}; \delta) \right\} \nearrow \inf \left\{ \mathbf{H}(\nu) : \nu \in \overline{A} \cap \mathfrak{M}(\overline{U}) \right\}$$

as $\delta \searrow 0$. $\quad \square$

**Remark 2.15.**

There are several advantages to the line of reasoning which we have taken. In the first place, (2.11) together with Varadhan's lemma (cf. Theorem 2.1.10 in [DS]) leads to

$$\lim_{\delta \searrow 0} \lim_{n \to \infty} \frac{1}{n} \log \left[ \frac{\int_{A_n(\delta)} \exp\left[-n\Phi(\mathbf{R}_n)\right] d\lambda^{\mathbf{Z}}}{\lambda^{\mathbf{Z}}(A_n(\delta))} \right]$$

$$= \sup \left\{ \langle \Phi, \mu \rangle - \mathbf{H}(\mu) : \mu \in \mathbf{M}_1^{\mathbf{S}}(\Omega) \right\}$$

for every $\Phi \in C(\mathbf{M}_1(\Omega); \mathbb{R})$. Secondly, it ought not be too difficult to extend the method presented here to cover higher dimensional lattices. Indeed, the analogue of Lemma 2.5 is more or less implicit in Ruelle's [R], and the appropriate extension of (2.14) presents no insurmountable difficulties (cf., for example, [C])[1].

## REFERENCES

[C] Comets, F., *Grandes déviations pour des champs de Gibbs sur $\mathbb{Z}^d$*, C.R. Acad. Sc. Paris, Série I **303** (1986), 511.

[DS] Deuschel, J.–D. and Stroock, D., *Large Deviations*, Pure and Appl Math. Series #137, Academic Press, Boston, 1989.

[DT] Dobrushin, R. and Tirozzi, B., *The Central Limit Theorem and the problem of equivalence of ensembles*, Comm. Math. Phys. **54** (1977), 173–192.

[G] Georgii, H., *Canonical Gibbs Measures*, Lec. Series in Math. #760, Springer–Verlag, N.Y.C., 1979.

[SZ] Stroock, D. and Zeitouni, O., *Microcanonical distributions, Gibbs' states, and the equivalence of ensembles*, to appear in the Festschrift volume for F. Spitzer.

M.I.T., RM. 2-272, CAMBRIDGE, MA 02140, U.S.A.

---

[1]Since the preparation of this note, the author, in collaboration with J.–D. Deuschel and H. Zessin, has successfully carried out this extension to finite dimensional lattices. The resulting article will appear in **Comm. Math. Phys.** under the title *Microcanonical Distributions for Lattice Gases.*

# Progress in Probability

*Editors*

Professor Thomas M. Liggett
Department of Mathematics
University of California
Los Angeles, CA 90024-1555

Professor Charles Newman
Courant Institute of
Mathematical Sciences
251 Mercer Street
New York, NY 10012

Professor Loren Pitt
Department of Mathematics
University of Virginia
Charlottesville, VA 22903-3199

*Progress in Probability* is designed for the publication of workshops, seminars and conference proceedings on all aspects of probability theory and stochastic processes, as well as their connections with and applications to other areas such as mathematical statistics and statistical physics. It acts as a companion series to *Probability and Its Applications,* a context for research level monographs and advanced graduate texts.

We encourage preparation of manuscripts in some form of TeX for delivery in camera-ready copy, which leads to rapid publication, or in electronic form for interfacing with laser printers or typesetters.

Proposals should be sent directly to the editors or to:
Birkhäuser Boston, 675 Massachusetts Avenue, Cambridge, MA 02139, U.S.A.